Praise for *The Dangerous Book for Boys*

'The perfect handbook for boys and dads.'

Daily Telegraph

'Full of tips on how to annoy your parents.'

Evening Standard

'An old-fashioned compendium of information on items such as making catapults and knot-tying...the end of the PlayStation may have been signalled.'

The Times

'Just William would be proud. A new book teaching boys old-fashioned risky pursuits...has become a surprise bestseller.'

Daily Mail

'If you want to know how to make crystals, master NATO's phonetic alphabet and build a workbench, look no further.'

Time Out

THE POCKET DANGEROUS BOOK FOR BOYS:

WONDERS OF THE WORLD

Many of the pieces in this edition have been selected from the much-loved *The Dangerous Book for Boys*. They are a collection of fascinating things about the world around us that every boy should know.

This edition is a perfect pocket format for readers to take everywhere with them.

Visit www.dangerousbook.co.uk for quizzes, games and more.

The Pocket
DANGEROUS
Book
for
Boys

WONDERS OF THE WORLD

Conn Iggulden & Hal Iggulden

HarperCollins*Publishers*

HarperCollins*Publishers*
77–85 Fulham Palace Road,
Hammersmith, London W6 8JB

www.harpercollins.co.uk

Published by HarperCollins*Publishers* 2008
1

DANGEROUS BOOK FOR BOYS™ and

Conn Iggulden and Hal Iggulden assert the moral right to
be identified as the authors of this work

A catalogue record for this book
is available from the British Library

ISBN-13: 978 0 00 728180 0

Set in Centennial CE 55 Roman by Envy Design Ltd

Printed and bound in Italy by
L.E.G.O. SpA - Vicenza

This book contains a number of activities which may be dangerous
if you do not follow the advice given. For those who are under 18, these activities should
be carried out under the supervision of an adult. The authors and the publishers exclude
all liability associated with this book to the extent permitted by law.

To all of those people who said 'You *have* to include…' until we had to avoid telling anyone else about the book for fear of the extra chapters. Particular thanks to Bernard Cornwell, whose advice helped us through a difficult time and Paul D'Urso, a good father and a good friend.

CONTENTS

INTRODUCTION

WHEN I WAS A LITTLE BOY, I was deeply influenced by the character of Billy the Cat in the *Beano* comic. He used to dress in black, climb on roofs and solve crimes. For months afterwards, I put on my P.E. plimsolls and climbed my parents' roof in Eastcote, looking for burglars I could foil in the act. Sadly, no one obliged me by committing a crime so I could leap on him.

Almost all my favourite childhood memories took place outside, from catching moths in matchboxes, to making fried bread from the old oil lamps that used to be part of builders' equipment. I will never forget the tree that fell down while I was still in it. As a father to four children, I know how important it is to give them the freedom to fall on their own heads once in a while. All the best adventures take place outside, even if burglars mysteriously fail to put in an appearance.

I read recently that if you decided to rid yourself of an unwanted child by leaving them outdoors to be kidnapped, you'd have to wait for approximately 200,000 years. I don't know about you, but I don't have that kind of patience. There's a lot more fear than actual danger outside the safe confines of a home and the truth is that heart disease kills more people than anything else. In short, it's less dangerous to let your kids out to play and keep fit.

I also read *Lord of the Flies* and, perhaps unusually, enjoyed the idea so much that I looked for opportunities to become savage. For one glorious summer, my brother and I fought constantly with a local gang. We had the worst of it by far until the day when the leader insulted a grown man and was badly injured. He was miles from home and the only adults he knew were our parents. I had the odd experience of walking home from school and seeing my mum driving past with my worst enemy, on the way to hospital.

Many years later, I was teaching in my first school when I came across the gang leader doing community service by cutting school hedges. I hadn't seen him since childhood and I stopped in surprise.

'Hello,' I said. 'Do you remember me?'

'You're Hal Iggulden, aren't you?' he replied.

I hesitated for a moment. 'Why, yes. Yes I am!' I said. My brother Hal still doesn't think this is as funny as I do.

Understanding the natural world is an odd pleasure. I can't quite explain why I take such satisfaction from being able to recognise Venus and Jupiter in the night sky, or knowing the name of butterflies and beetles when I come across them. It doesn't make me a better person, or more attractive to women, I'm *reasonably* certain. It doesn't mean I can do backflips, another childhood ambition that ended with a trip to the emergency ward. However, there are things I feel I should know, that every man, and especially every father, should know. Boys soak up knowledge if their dad seems to enjoy reading. Perhaps some of them will use it working for N.A.S.A., or becoming a lepidopterist. I used to think that was a terrible disease, by the way, but I know better now. Even if it isn't used, even if the name of a beetle is no earthly use whatsoever, it's still valuable because it's fun to know things. That's how we're made, for some reason.

When I am researching historical fiction, I sometimes come across people who are experts in a very limited field – horse endurance racing, say, or ancient sword making. That sort of knowledge earns respect, for without it, we are little more than passive consumers. Knowledge makes us human, gives us control and leads to exciting discoveries like carbon fibre, new elements and electric cars. In terms of technology and knowledge, this is very much the world I hoped for when I was climbing those roofs as the 'Eastcote Cat'. My father used to say that there are two educations – the one at school and the far more important one that lasts for the rest of your life. I hope to enjoy my second one until my second childhood begins. In fact, I hope that one is as much fun as the first.

Conn Iggulden

THE SEVEN WONDERS OF
THE ANCIENT WORLD

THE FAMOUS SEVEN WONDERS of the ancient world were: the Great Pyramid of Cheops at Giza, the Hanging Gardens of Babylon, the Temple of Artemis at Ephesus, the Mausoleum at Halicarnassus, the Colossus of Rhodes, the Statue of Zeus at Olympia and the Pharos Lighthouse at Alexandria. Only the pyramid at Giza survives to the modern day.

I. THE GREAT PYRAMID

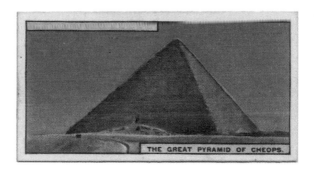

THE GREAT PYRAMID OF CHEOPS.

The Great Pyramid is the largest tomb ever built, created for the 4th Dynasty Egyptian pharaoh Khufu (2898–2875 BC), though he is better known by the Greek form of his name, Cheops.

It is one of the three great pyramids at Giza near Cairo, the other two being constructed for the pharaohs Menkaure and Khafre. The largest, for Cheops, was the tallest structure on earth for more than four thousand years, until the nineteenth century AD. Though the capstone was removed at some point, it would have stood at 481 ft (146.5 m) high.

The base is perfectly square – a feat of astonishing accuracy considering the sheer size of it. Each side of the base is 755 ft 8 in (231 m) long and each side slopes at 51 degrees, 51 minutes. It is composed of two million blocks of stone, *each one* weighing more than two tons. They fit together so well that not even a knife blade can be slid between them.

2. The Hanging Gardens of Babylon

The Hanging Gardens of Babylon were built in what is now modern-day Iraq, on the banks of the river Euphrates. They were created by King Nebuchadnezzar for his queen between the seventh and sixth centuries BC.

Famously, they employed complex hydraulic systems to raise thousands of gallons from the river and keep the gardens blooming. We can only guess at the exact method, but an Archimedean screw may have been employed.

THE HANGING GARDENS OF BABYLON.

3. The Temple of Artemis (Diana) at Ephesus

The Temple of Artemis (Diana) at Ephesus in what is modern-day Turkey is said to have awed Alexander the Great with its extraordinary beauty, though the citizens refused his offer to bear the cost of a restoration. Originally built in the sixth century BC, the temple was destroyed and rebuilt on more than one occasion, though the most famous was the night of Alexander's birth, when a man named Herostratus burned it so that his name would be remembered – one of the greatest acts of vandalism of all time. It finally fell into ruin around the third century AD.

TEMPLE OF DIANA AT EPHESUS.

THE SEVEN WONDERS OF THE ANCIENT WORLD

4. The Mausoleum at Halicarnassus

The Mausoleum at Halicarnassus was created for King Mausolus of Persia, who ruled from 377 to 353 BC. Halicarnassus is now the city of Bodrum in Turkey. On top of the rectangular tomb chamber, thirty-six columns supported a stepped pyramid crowned by statues of Mausolus and his wife (and sister) Artemisia in a chariot, reaching a height of approximately 140 ft or 42.5 m. It was destroyed in 1522 when crusading Knights of St John used the stone to build a castle that still stands today. The polished marble blocks of the tomb are visible in the walls. From Mausolus, we have the word 'mausoleum', meaning an ornate tomb.

MAUSOLEUM AT HALICARNASSUS

5. The Statue of Zeus at Olympia

The Statue of Zeus at Olympia is also lost to the modern world. Only images on coins and descriptions survive to tell us why the statue was considered so astonishing in the fifth century BC.

Olympia was the site of the ancient Olympic games – giving us the word. The site was sacred to Zeus, and Phidias of Athens was commissioned to carve the statue.

The statue was of wood layered in gold for the cloth and ivory sheets for the flesh. In his right hand stood the winged figure of the goddess Victory (Nike), made of ivory and gold. In his left, he held a sceptre made of gold, with an eagle perched on the end.

STATUE OF ZEUS AT OLYMPIA.

The Roman emperor Caligula tried to transfer the statue to Rome in the first century AD, but the scaffolding collapsed under the weight and the attempt was abandoned. Later on, the statue was moved to Constantinople and remained there until it was destroyed by fire in the fifth century.

6. The Colossus of Rhodes in Greece

The Colossus of Rhodes in Greece is perhaps the most famous of the seven ancient wonders. It was a statue of Helios, over a hundred feet (30 m) high.

It did not actually stand across the harbour, but instead rested on a promontory, looking out over the Aegean Sea. The base was white marble and the statue was built slowly upwards, strengthened with iron and stone as the bronze pieces were added. It took twelve years and was finished around 280 BC, quickly becoming famous. An earthquake proved

THE COLOSSUS OF RHODES.

disastrous for the statue fifty years later. It broke at the knee and crashed to the earth to lie there for eight hundred years before invading Arabs sold it.

7. THE PHAROS LIGHTHOUSE AT ALEXANDRIA

The Pharos Lighthouse at Alexandria was built by the architect Sostratus of Cnidus for the Greco-Egyptian king Ptolemy Philadelphus (285–247 BC).

THE PHAROS OF ALEXANDRIA.

Ptolemy's ancestor had been one of Alexander the Great's generals. His most famous descendant is Cleopatra who was the first of her Greek line actually to speak Egyptian.

When Julius Caesar arrived in Alexandria, he would have passed by the great lighthouse on Pharos island. Its light was said to be visible for 35 miles (55 km) out to sea. Its exact height is unknown, but to have shed visible light to that distance, it must have been between 400 and 600 feet high (121–182 m).

It was so famous that, even today, the word for lighthouse in Spanish and Italian is 'faro'. French also uses the same root, with 'phare'.

As you can see, even the greatest wonders can be lost or broken by the passage of millennia. Perhaps the true wonder is the fact that we build them, reaching always for something greater than ourselves

QUESTIONS ABOUT
THE WORLD – PART ONE

—◆—

1. Why is a summer day longer than a winter day?
2. Why is it hotter at the Equator?
3. What is a vacuum?
4. What is latitude and longitude?
5. How do you tell the age of a tree?

1. WHY IS A SUMMER DAY LONGER THAN A WINTER DAY?

In Australia, the shortest day is 21 June, and the longest falls on 21 December. In the northern hemisphere, 21 June is midsummer and midwinter falls on 21 December. Christmas in Australia is a time for barbecues on the beach.

Although the North Pole points approximately at the star Polaris, the Earth's axis is tilted twenty-three and a half degrees in respect to the path it takes around our sun.

While the northern hemisphere leans towards the sun, more direct sunlight reaches us. We call this period summer. 21 June is the day when the North Pole points directly towards the sun, and the tilt is at maximum. The days are longest then as most of the northern hemisphere is exposed. Down in the south, the days are shortest as the Earth itself blocks light from reaching the shivering inhabitants.

As the Earth moves around the sun, the tilt remains the same. The autumnal equinox (22 or 23 Sept) is the day when day and night are of equal length – twelve hours each, just as they are on the vernal equinox in spring on 20 March. 'Equinox' comes from the Latin for 'equal' and 'night'.

When the northern hemisphere leans away from the sun, less light reaches the surface. This is autumn for us, and eventually winter. Longer days come to the southern hemisphere as shorter days come to the north. The summer solstice of 21 June is also the moment when the sun is highest in the sky.

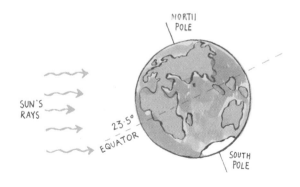

The Earth is actually closer to the sun in January rather than June. It's not the distance – it's the tilt.

The best way to demonstrate this is by holding one hand up as a fist and the other as a flat palm representing the Earth's tilt. As your palm moves around the fist, you should see how the tilt creates the seasons and why they are reversed in the southern hemisphere. Be thankful that we have them. One long summer or one long winter would not support life.

At the midsummer and midwinter solstices, the conditions can become very peculiar indeed. The summer sun will not set for six months at the North and South Poles, but when it does set, it does not rise for another six. Northern countries such as Finland also experience the 'midnight sun' effect.

2. WHY IS IT HOTTER AT THE EQUATOR?

There are two reasons why the Equator is hotter than the rest of the planet. Strangely enough, the fact that it is physically closer to the sun than, say, the North Pole is not relevant. The main reason is that the Earth curves less in the equatorial region. The same amount of sunlight is spread over a smaller area. This can be clearly seen in the diagram opposite.

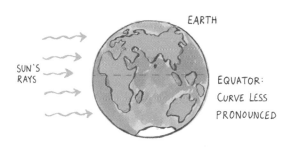

EARTH

SUN'S
RAYS

EQUATOR:
CURVE LESS
PRONOUNCED

Also, the sun's rays have to pass through less atmosphere to reach the equatorial band – and so retain more of their heat.

3. What is a vacuum?

A perfect vacuum is a space with absolutely nothing in it – no air, no matter of any kind. Like the temperature of absolute zero (−273.15 °C/0 Kelvin), it exists only in theory. The light bulbs in your home have a 'partial vacuum', with most of the air taken out as part of the manufacturing process. Without that partial vacuum, the filament would burn far faster, as air contains oxygen.

The classic science experiment to show one quality of a vacuum is to put a ticking clock inside a bell jar and expel the air with a pump. Quite quickly, the sound becomes inaudible: without air molecules to carry sound vibrations, there can be no sound. That is why in space, no one can hear you scream!

4. What is latitude and longitude?

The Earth is a globe. The system of latitude and longitude is a man-made system for identifying a location anywhere on the surface.

PARALLELS OF LATITUDE

Latitude takes the Equator as a line of zero. If you cut the world in half at that point, you would have a horizontal plate. The centre point of that plate is at ninety degrees to the Poles above and below it.

Latitude is not measured in miles but in the degrees between ninety and zero in both hemispheres. London, for example is at 51° latitude north.The curve representing the ninety-degree change is split into imaginary lines called 'parallels' – because they are all parallel to each other and the Equator.

PARALLELS of LATITUDE

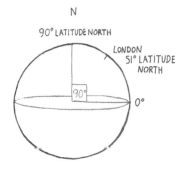

With something as large as the Earth, even a single degree can be unwieldy. For both longitude and latitude, each degree is split into sixty 'minutes of arc'. Each minute of arc is split into a further sixty 'seconds of arc'. The symbols for these are:

Degrees: °   Minutes: '   Seconds: "

With something as large as a city, the first two numbers would suffice. London would be 51° 32′ N, for example. The location of a particular house would need that third number, as well as a longitude coordinate.

There is an element of luck in the fact that a latitude degree turned out to be almost exactly sixty nautical miles – making a minute of latitude conveniently close to one nautical mile, which is 6,000 feet (1852 metres).

The *longitude* of London is zero, which brings us neatly into longitude.

Longitude is a series of 360 imaginary lines stretching from Pole to Pole. London is zero and 180 degrees stretch to the west or east.

If the world turns a full circle in a day, that is 360 degrees. 360 divided by 24 = 15 degrees turn every hour. We call the fifteen-degree lines 'meridians'. ('Meridian' means 'noon', so there are twenty-four noon points around the planet.)

Now, this is how it worked. On board your ship in the middle of nowhere, you took a noon sighting – that is, took note of the time as the sun passed its highest point in the sky. You could use a sextant and a knowledge of trigonometry to check the angle. If you were at noon and your ship's clock told you Greenwich was at nine in the morning, you would have travelled three meridian lines east or west – which one depending on your compass and watching the sun rise and set. You would be at longitude +/– 45°, in fact.

Having a clock that could keep the accurate time of Greenwich even while being tossed and turned on a ship was obviously crucial for this calculation. John Harrison, a clock maker from Yorkshire, created a timepiece called H4 in 1759 that was finally reliable enough to be used.

All that was left was to choose the Prime Meridian, or zero-degree point of longitude. For some time it looked as if Paris might be a possibility, but trade ships in London took their time from the Greenwich clock at Flamstead House, where a time ball would drop to mark 1 p.m. each day. Ship chronometers were set by it and Greenwich time became the standard. In 1884 a Washington conference of twenty-five nations formalised the arrangement. If you go to Greenwich today, you can stand on a brass line that separates the west from the east.

On the opposite side of the world, the two hemispheres meet at the International Date Line in the Pacific Ocean. It's called the International Date Line because we've all agreed to change the date when we cross it. Otherwise, you could travel west from Greenwich, back to 11 a.m., 10 a.m., 9 a.m., all the way round the planet until you arrived the day before. Obviously this is not possible, and so crossing the line going west would add a day to the date. Complex? Well, yes, a little, but this is the world and the systems we made to control it.

Like latitude, longitude is broken down into a three-figure location of degrees, minutes and seconds. Common practice puts the latitude figures first, but it's always given away by the North or South letter, so they can't really be confused. A full six-figure location will look something like these:

38° 53′ 23″ N, 77° 00; 27″ W	Washington DC
39° 17′ 00″ N, 22° 23; 00″ E	Pharsalus, Greece, where Julius Caesar beat Pompey and ended the civil war.
39° 57′ 00″ N, 26° 15′ 00″ E	Troy

6. How do you tell the age of a tree?

You cut it down and count the rings. For each year of growth, a dark and a light ring of new wood is created. The two bands together are known as the 'annual ring'. The lighter part is formed in spring and early summer when the wood cells are bigger and have thinner walls which look lighter. In autumn and winter, trees produce smaller cells with thicker walls which look darker. They vary in width depending on growing conditions, so a tree stump can be a climate record for the life of the tree – sometimes even centuries. The age of a tree, therefore, can be told by counting the annual rings.

INSECTS AND SPIDERS

'INSECT' AS A WORD is from the Latin, meaning 'cut into' or 'segmented'. An insect is any creature with a head, a thorax, an abdomen and six legs. They usually have an exoskeleton – protective plating on the outside. They are by far the largest class in the animal kingdom. There are hundreds or even thousands of different species to be found in any field or stretch of open water in the country. They are part of fantastically complex ecosystems and in a single village pond a hundred thousand lives can come into existence, fly and perish, sometimes even in a single day. Their variety is astounding and their lives can be endlessly fascinating. Here are some of the ones you might find near where you live.

GRASSHOPPERS (ORTHOPTERA)

Although, with tiny differences, there are more than two dozen varieties of **grasshopper**, they can be put into two main groups: long-horns (*Locustidae*) and short-horns (*Acrididae*). Both make the familiar rhythmic creaking noise on sunny days, though short-horned varieties are

A Meadow Grasshopper   *Cricket*

much more common. The long-horned grasshopper can be as much as five times larger than their cousins and are capable of flight, though usually only in very short bursts. When they are stationary, they are practically invisible. To find them, walk very slowly through long grass, the longer the better. In the summer, you will see small darting specks of small meadow grass-hoppers leaping away from you. They are usually bright green, but can also be found in brown or grey. If you are lucky, you will see a larger long-horned one. By all means try to catch the small grasshoppers, but the long-horns are always damaged when they are caught by hand.

Crickets are far less common in this country, though there are three varieties: the field cricket, the house cricket and the mole cricket, which spends most of its time underground. The British field cricket is an endangered species and came close to dying out in the 1980s. They have been bred and reintroduced to the wild since then, but sightings are still relatively rare. They are more common in France, Italy and many warmer countries, however.

EARWIGS (DERMAPTERA)

These are so common that it might seem odd to put them here – the reason is merely to say that **earwigs** are completely harmless. They are nocturnal insects, with one flying variety. The fierce-looking clippers are for holding, not killing. The female cares for and feeds her young, after laying eggs in a tiny nest dug with the male.

Earwig

Mayfly

Mayflies (Ephemeroptera)

The most fascinating thing about **mayflies** is their life cycle. They live for only a few hours, emerging from a chrysalis without even a mouth to feed. The final brief flight of its life comes after a much longer period as a nymph grub underwater. More than one poet or writer has seen within the story of the mayfly a metaphor for our own short time in the sun. A lifetime is just a matter of scale.

The mayfly lives only to mate and despite the apparent fragility of such a system, they have been found preserved in fossil form in Paleozoic era rocks, three hundred and fifty million years ago – before even the dinosaurs!

Dragonflies and Damselflies (Odonata)

Another harmless and beautiful group of insects. Both **dragonflies** and **damselflies** belong to the order Odonata, meaning 'toothed jaw'. Their lower jaws are serrated, which may explain the name. Even large ones are incapable of breaking human skin, however. Having four wings makes them wonderfully alien, though it is their bright colours that catch the eye in summer. In addition, they consume gnats and mosquitoes, so are a very welcome presence in a garden.

Dragonfly   *Damselfly*

As with the mayfly, the grub stage hatches underwater, then crawls up a reed or aquatic plant until it reaches air. The skin hardens and splits and a dragonfly struggles out of its old carcase, born anew. Damselflies are a suborder (*Zygoptera*), with four wings of roughly equal size. In comparison, dragonflies (*Anisoptera*) have hind wings that are shorter and broader than the forewings. Overall, there are more than forty different types, though by far the most common are the blue-tailed and common blue 'needle' damselflies, found in eastern and southern England in particular.

All dragonflies have excellent eyesight and flying skills – they need them to survive fast attacks from birds and slower ones from frogs if they come down to water to lay eggs or to drink.

They are strictly summer insects and do not survive cold weather. Wet weather too can starve them as neither dragonflies nor their prey fly in the rain.

Water Surface Insects

The **pond skater** (*Gerris lacustris*) uses the surface tension of water to scull itself along without getting wet. Its already tiny weight is spread on long legs, as can be seen in the image here. The front legs row it along at an astonishing speed for its size.

The **water boatman** (*Notonecta glauca*) rows along on its back, again at a fair clip for such a tiny insect. Unlike the pond-skater it is carnivorous. Neither of these poses any danger to us, they are simply strange and fascinating members of the insect world.

Pond Skater

Water Boatman

Butterflies and Moths (Lepidoptera)

There are many hundreds of species to be seen in the British countryside. Moths fly by day and night, though butterflies restrict themselves to daylight hours only. Another difference is that butterflies generally raise their wings erect when at rest and have little buds or clubs at the end of their antennae. Although most moths are brown or black in colour, some can be as bright as butterflies.

Here are a few of the ones you are most likely to see in fields and gardens.

Gatekeeper or **H**edge Brown (*Pyronia tithonus*).

The **Red Admiral** (*Vanessa atalanta*) is one of the best known, though it is often confused with the similar Painted Lady. The Red Admiral is an attractive flutterer found wherever there are wild nettles, fruit or hops. The Latin name is from Atalanta, a Greek girl who refused to

Gatekeeper

Red Admiral

Peacock

Large White   *Small Blue*

marry unless a man could beat her in a foot race. As suitors chased her, so children still chase the Red Admiral today.

The **Peacock Butterfly** (*Inachis io*) is named for the eyes on its wings. Like the Red Admiral, they eat and lay their eggs on stinging nettles. They can be seen in spring and summer.

The **Cabbage White** (*Pieris rapae*) is a bit of a pest to gardeners, as the name suggests. Nonetheless, it is one of the most common British butterflies.

The **Small Blue** (*Cupido minimus*) is the smallest British butterfly and is no larger than the tip of a finger. It is found on moors and grasslands in summer and produces a similarly small brown caterpillar from its eggs that has fine brown bristles and a glossy black head.

MOTHS

Moths are a common sight whenever a window is left open at night. Their variety is immense. In fact, of around 130,000 species of Lepidoptera in the world, moths account for 110,000 of them. Famously, their senses are confused by bright light and they can spend many unhappy hours bumping against bulbs. In previous generations, the light would have come from a flame and the moth would be drawn to it and then burned. The metaphor is obvious when considering anything else lured to its own destruction.

Like butterflies, they spend time as caterpillars, emerging as adults from a chrysalis. Some are brightly coloured and fly by day; only the lack of clubbed antennae can show that you are looking at a moth rather than a butterfly.

Six-spot Burnet

INSECTS AND SPIDERS

The **Six-spot Burnet** or **Ten O'clock Riser** (*Zygaena filipendulae*). Can be found in Scotland in July and August.

Finally, one of the most useful moths in the world is *Bombyx mori*. The moth is practically unknown, but its caterpillar larvae are silkworms and still produce all the world's natural silk, unwound from their cocoons. They have been bred in China for five *thousand* years.

BEETLES (COLEOPTERA)

Beetles are insects with a hard carapace protecting wings. Many are scavengers and play a vital role in consuming dead animals and birds. There are around 3,600 different species in this country.

The **Dor Beetle** or 'Dumble-Dor' (*Geotrupes stercorarius*) shown here, buries cow dung as a food source. It is benign and relatively common. Other species are positively destructive, however, such as the brown **Deathwatch Beetle** (*Xestobium rufovillosum*) that bores holes in wood and can destroy old beams and buildings.

Glow-worms (*Lampyris noctiluca*) are not worms at all. They too are beetles. Sightings are quite rare. The males fly, but their light is very dim. The females are flightless, but give off a much brighter yellow-green light that can be

Dor Beetle

Glow-worms

seen in country hedges at dusk in May. One grisly fact about the glow-worm is that its larvae seek out inhabited snail shells when they hatch, feeding on the defenceless snails within.

Ladybirds (*Coccinelidae*) are a very familiar beetle and can be found in any grassy meadow. They eat greenfly and are welcome in any garden. If annoyed, they eject an unpleasant-tasting fluid as a defence, just as a grass snake does. (If you ever pick up a grass snake, be prepared for a cupful of the worst-smelling filth you have ever experienced. One of the authors was caught unawares trying this and the smell lingered for days despite endless hand-washing in powerful detergent.)

The **Stag Beetle** (*Lucanus cervus*) is not particularly uncommon, though the authors have only ever seen one. We kept him in a matchbox until he somehow escaped. As with earwigs, the horns of Britain's largest beetle are

Ladybirds   *Stag Beetle*

completely harmless. Males cannot be kept with other
males as they will fight and damage or even kill each other.
Also, pairs must be kept apart after mating, or they will bite
each other's legs off. The life of a Stag Beetle is not an easy
one! They can be bred in captivity, but the pupae are very
easy to damage and should not be touched by bare skin.

DEES AND WASPS

Bees are fascinating insects – and extremely unlikely to
sting unless you make them afraid. If you sit on one, it will
sting you, but in the circumstances, who could blame it?
Otherwise, they are harmless and, of course, they produce
delicious honey. The **Bumble Bee** (*Bombus terrestris*,
sometimes called the Humble Bee) can be seen bumbling
around looking for nectar in the summer, though it is less

common than the common Honey Bee or Hive Bee (*Apis mellifera*). Very few Honey Bees are wild in Britain any more. Their lives could fill a chapter on their own, but the main types are workers, drones and queens. The drones live only for a single season, while the queen lives three or four years.

Wasps are almost universally disliked. The **Common Wasp** (*Vespula Vulgaris*) comes in varieties of non-reproductive workers, males and queens, the queens being larger than the rest. They can be aggressive if attacked and will sting with very little provocation. If they are trapped under clothing, they can sting more than once.

The **Hornet wasp** (*Vespa crabro)* is much larger than

Bumble Bee

Common Wasp   *Hornet*

the common variety and has brown bands rather than black. Thankfully, they are not common.

The pain-causing chemical injected by a bee or wasp sting is called 'melittin'. A bee sting usually rips out the whole sting apparatus from the bee in the process, wounding it fatally. Sadly, the wasp has no such handicap and can fly away happily after stinging.

ANTS (FORMICIDAE)

There are around fifty different species of ants in Britain, though most are harmless. Black or yellow ants of any size, whether winged or not, cannot harm humans. **Black Wood Ants** (*Formica rufa*) can eject an unpleasant spray of formic acid, however, which smells like bitter vinegar. Anyone who has ever sat down on a red ant nest will know how painful their bites can be. **Red Ants** (*Myrmica ruginodis*) are aggressive and unfortunately seem to enjoy the garden habitat as much as their black cousins (*Lasius niger*).

FLIES AND MOSQUITOES

Bluebottles (*Calliphora vomitoria*) and **Greenbottles** (*Lucilia caesar*) lay eggs which hatch into maggots. Apart from being useful for fishing, they spread dirt and disease and should be kept away from food if at all possible. They are attracted to rotting meat, household rubbish and excrement in any form. There really isn't anything pleasant to say about them.

Hoverflies (*Syrphidae*) look a little like small wasps, but

Bluebottle

Hoverfly

Horsefly

Midge

Gnat

can be recognised by their hovering, darting motion and are harmless. **Horseflies** (*Tabanidae*), on the other hand, are an absolute menace, as one of the authors found out on a Scottish hillside once. Their bite leaves a tiny bleeding hole. Both authors have been subject to the attention of **Midges** (*Ceratopogonidae*) in Scotland. They leave itchy red marks on the skin and swarm around water in extraordinary numbers.

The **Common Gnat** (*Culex pipiens*) is very similar-looking to the more dangerous **Malarial Mosquito** (*Anopheles maculipennis*). Both are members of the same family and females from both species feed off humans if they get the chance, making a characteristic whining sound just as you are trying to get to sleep. In countries like Italy and France, they are a serious pest and whole areas have to be sprayed regularly. Malaria carried by the Anopheles Mosquito is still a terrible killer in parts of Africa. They are not common in Britain.

WOODLICE

The **Pill Woodlouse** (*Armadillidium vulgare*) is capable of rolling itself into a tight ball, hence the name. They are harmlessly amusing creatures and less common than the blue-grey **Common Woodlouse** (*Porcellio scaber*), which can be found wherever there is rotting wood or dampness.

SPIDERS (ARACHNAE)

Spiders are not insects. They have eight legs rather than six, have only two sections to their bodies and have eight single eyes instead of two compound ones. Britain is extremely lucky in the six hundred varieties of spiders found naturally within her borders. None are dangerous. In comparison with many other countries, a small child can be allowed to wander barefoot in, say, Wales, without worrying that they will be bitten or even killed.

The common **House Spider** (*Tegenaria atrica*) is completely harmless, though it can be quite large in country settings and moves worryingly quickly across the floor when it senses danger.

Another common sight in wooden sheds everywhere is the **Garden Spider** (*Arachneus diadematus*). Again, these

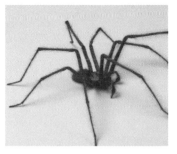

House Spider

Garden Spider

can grow quite large with a good supply of flies and smaller spiders. It makes funnel-style webs and can be tempted out by touching a leaf or pencil to the edge of one.

There are many other species of spider and many thousands more insects with different and interesting lives and habits. The more you learn about insects, the more you understand what an incredibly complex world this actually is.

THE MOON

THROUGH ALL HUMAN HISTORY, the moon has drawn the gaze upwards. It was there in ancient myths; it was the light for a million romantic evenings – and it was our first stepping stone to the darkness beyond it. The gravity well of earth is crushingly powerful. Without the moon as a launching stage, regular space flight may never be possible. While it sails above, we can dream of lunar bases and leaving the earth behind.

The first landing on the moon was on 20 July 1969, one date *everyone* should know. It is the only object in space that we have visited, after all. The *Apollo 11* spacecraft reached the moon and fired braking rockets to take up orbit around it. Neil Armstrong and Edwin 'Buzz' Aldrin descended to the surface in a landing module named 'Eagle'. Michael Collins remained in the command module. After announcing to the watching earth that 'the Eagle has landed', Armstrong stepped out onto the surface of the moon.

There have been many momentous events in our history, from Caesar crossing the Rubicon to the first use of an atomic bomb, but having a human being set foot on another, stranger soil may be the most extraordinary.

Armstrong's first words were, 'This is one small step for man, one giant leap for mankind.' Famously, he had intended to say 'a man'. Without the 'a', he seemed to repeat himself.

The two men spent twenty-one hours on the surface and brought back forty-six pounds of moon rock. The moon has no atmosphere – and therefore no protection from meteorites. Its surface has been battered and melted by these strikes over billions of years, resulting in a soil called a 'regolith' – made of dust, rock and tiny beads of glass that are slippery underfoot.

The *Apollo 11* landing was the first of six successful landing missions during the twentieth century. In sequence, they are: *Apollo 11*, *12*, *14*, *15*, *16* and *17*, ending in December 1972. *Apollo 13* suffered technical problems and had to return to earth without landing on the moon. There will be others. An unmanned probe named *Lunar Prospector* found ice in 1998 at both moon poles – one of the most important requirements of a future colony!

THE PHASES OF THE MOON – AS SEEN FROM EARTH

The phases of the moon are such a part of our world that they should be common knowledge.

1. This is a new moon. The moon is between the earth and the sun and shows no light. This position creates the strong vernal tides on earth.

2. A waxing (growing) crescent. This use of the word 'waxing' is now almost completely restricted to describing the phases of the moon. As the moon moves on its cycle, we see the sunlight reflecting on its surface. The crescent will grow as it moves around the earth.

3. First quarter moon. Quarter of the way around the earth, one clear half of the moon is visible.

4. Waxing gibbous. The word 'gibbous' is another one that tends to be used only in descriptions of the moon. It means convex, or bending outwards. This is a good time to take pictures of the moon. Surprisingly sharp images can be made by the simple action of putting a camera up against the lens of a telescope on a tripod.

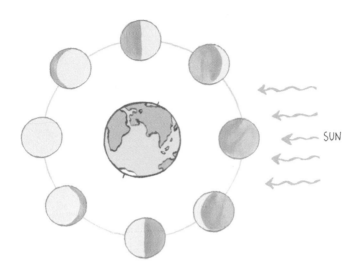

5. Full moon – also a time of strong tides, as both the sun and moon pull at the earth's oceans.

6. Waning moon – beginning the path back to the new moon.

7. Last quarter, with a perfect half of the moon again visible.

8. Waning crescent.

MOON FACTS

1. Distance from earth: because of an elliptical orbit, this varies, but on average is 240,000 miles (386,000 km).

2. Gravity: about 1/6 of earth.

3. Day length: 27.3 earth days.

4. Time to orbit earth in relation to a fixed star (sidereal month): 27.3 earth days.

5. Time in relation to the sun (new moon to new moon/synodic month): 29.5 days.

6. Because it takes 27.3 days to orbit earth *and* turn on its own axis, we always see the same face. (See Questions About the World – Part Two.) However, there is no dark side in the sense of lacking light. Like earth, there is a night and day side, but both receive light during the cycle. The 'dark side' of the moon just doesn't exist!

7. The moon has no atmosphere, which means no wind, so Neil Armstrong's original footprint will still be there

exactly as it was in 1969 – unless Buzz Aldrin or one of the others scuffed it over.

8. Daytime temperatures can reach up to 273 °F (134 °C). That is almost three times as hot as the Sahara Desert on earth. Night-time temperatures can be as low as –243 ° F (–152 °C). Needless to say, human beings cannot survive such an extreme range without a great deal of protection.

9. The American flag planted by the *Apollo 11* astronauts had to be made out of metal. Without an atmosphere, a cloth flag would have hung straight down.

10. The moon is silent. Without air or some other medium, sound waves cannot travel.

11. We owe many of the beautifully named parts of the moon to Galileo. It was he who thought he saw oceans on the moon in 1609, giving us Mare Tranquillitas (Sea of Tranquillity), Mare Nectaris (Sea of Nectar), Mare Imbrium (Sea of Showers), Mare Serenitatis (Sea of Serenity) and many more. Sadly, they are dry depressions and not the great oceans of his imagination.

1. Tycho Crater.
2. Mare Nectaris (Sea of Nectar).
3. Mare Fecunditatis (Sea of Fertility).
4. Mare Crisium (Sea of Crises).

5. Landing point of *Apollo 11*, on south-west edge of Mare Tranquillitas (Sea of Tranquillity).
6. Mare Serenitatis (Sea of Serenity).
7. Mare Imbrium (Sea of Showers).
8. Mare Frigoris (Sea of Cold).
9. Mare Nubium (Sea of Clouds).
10. Copernicus Crater.

LUNAR AND SOLAR ECLIPSES

Every month, at full moon, the earth goes between the sun and the moon. However, the exact line-up required for a lunar eclipse is not so common. Usually, the moon's tilted orbit takes it out of alignment. At most, there are only two or three full eclipses of the moon each year. You might expect to see around forty in a lifetime.

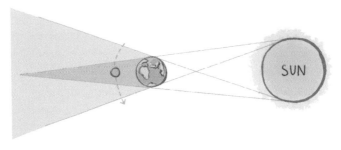

THE MOON

If you were standing on the moon at the time of a lunar eclipse, you would see the earth slowly blotting out the sun. In a lunar eclipse, the earth's shadow falls across the moon, though red sunlight scattered from the earth's atmosphere sometimes means a red moon can still be seen. Given the relative sizes, the earth's cone of shadow completely covers the moon and the eclipse can be seen from anywhere on the earth's surface.

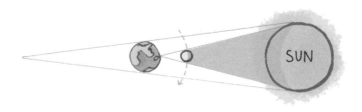

A solar eclipse is a far rarer event. In these, the full eclipse can only be seen along a narrow track, never more than about 200 miles wide. Obviously, these can only take place at the time of a new moon, when the shadow of the moon falls onto earth.

There are two kinds of solar eclipse, 'annular' and 'total'. Annular eclipses are about twice as common and far less impressive. They occur when the moon is too far from earth to block the sun completely. The sky will not darken as completely and a bright ring will still be visible around the moon. Annular means 'ring-shaped'.

A total solar eclipse is well worth travelling to see. It is one of the marvels of the natural world. First, a tiny bite appears in the ring of the sun, which deepens until the sun becomes a crescent and the day darkens towards an eerie twilight. The corona of the sun can then be seen around the black disc. The temperature drops and birds often return to trees to roost. Then the light begins to reappear and the world as we know it returns.

DINOSAURS

$\approx\!\!\!\!\times\!\!\!\!\approx$

THE TERM 'DINOSAUR' means 'terrible lizard', coined by a British scientist, Richard Owen, in 1842. These reptiles roamed the earth for over a hundred and fifty million years, then mysteriously died out. They varied from fierce killers to gentle plant eaters.

The largest dinosaurs were also the largest land animals ever to have existed. In 1907, the immense bones of a Brachiosaurus were discovered in east Africa. When alive, the animal would have been 75 ft (23 m) long and weighed between fifty and ninety tons. Its shoulder height would have been 21 ft (6.4 m) off the ground. These giants rivalled the largest whales in our present present-day oceans. In comparison, the largest living land animals today, elephants, weigh only five tons!

The Age of the Dinosaurs

The age of the dinosaurs is known as the Mesozoic era. This stretched from 248 to 65 million years ago. It divides into three separate time spans: the Triassic, the Jurassic and the Cretaceous. At the start of the Mesozoic era all the continents of today's earth were joined together in one super continent – Pangaea. This was

ERAS	PERIODS
Cenozoic	Quaternary
	Tertiary
Mesozoic	Cretaceous
	Jurassic
	Triassic
Paleozoic	Permian
	Pennsylvanian
	Mississipian
	Devonian
	Silurian
	Ordovician
	Cambrian

0
50
100
150
200
250
300
350
400
450
500
Millions of
years ago

surrounded by a massive ocean called Panthalassa. These names sound quite impressive until you realise they mean 'the whole earth' and 'the whole sea'. The German geophysicist, Alfred Wegener first came up with the theory of moving tectonic plates, or 'continental drift', in 1912. He examined similarities in rocks found as far apart as Brazil and southern Africa and realised they came from a single landmass.

Brachiosaurus – the 'Arm lizard'.

The Triassic world saw the first small dinosaurs, walking on their hind legs. This period lasted from 248 to 206 million years ago. Over millions of years Pangaea split into continents and drifted apart. After separation, different groups of dinosaurs evolved on each continent during the Jurassic period from 206 to 144 million years ago. This was the era of the giants. Huge herbivorous dinosaurs roamed in forests and grassland that covered entire continents.

The continental 'plates' are still moving today. In fact, wherever an area is prone to earthquakes or volcanoes, the cause is almost always one plate pushing against another, sometimes deep under the sea. The vast mountain ranges of the Andes and the Rockies were formed in this way.

The Cretaceous period lasted from 144 to 65 million years ago. This age included armoured plant eaters like Triceratops, browsers like Hadrosaur and huge meat eaters like the Tyrannosaurus Rex.

The seas too were filled with predators and prey that were very different from the inhabitants of today – except for sharks, oddly enough, who seem to have reached a perfect state of evolution and then stuck there for millions of years. Crocodiles are another example of a dinosaur that survived to the modern world. Modern crocodiles and alligators are smaller than their prehistoric cousins, but essentially the same animals. A crocodile from the Cretaceous period would have stretched to 49 ft (15 m)!

The World of the Dinosaurs

The dinosaurs' world was hot and tropical and dinosaurs of many shapes and sizes roamed pre-historic earth. One of the most interesting things about studying dinosaurs is seeing how evolution took a different path before the slate was wiped clean in 65 million years BC. Carnivores developed into efficient killing machines, while their prey either grew faster, or more heavily armoured as the eras progressed – the original arms race, in fact. Huge herbivores could nibble leaves from tree tops as tall as a five-storey building. The largest were so immense that nothing dared attack a healthy adult, especially if they moved in herds. The herbivores must have eaten huge amounts of greenery each day to fill their massive bodies – with stones, perhaps, to grind up the food in their stomachs.

As well as the giants, the age of dinosaurs overshadowed a smaller world of predators and prey. Compsognathus was only about the size of a modern house cat. We know it ate even smaller lizards as one has been found preserved in a Compsognathus stomach cavity.

The fastest group of dinosaurs were probably the two-legged ornithomimids – the 'ostrich mimics'. It is always difficult to guess at speed from a fossil record alone, but with longer legs than Compsognathus, they may have been able to run as quickly as a modern galloping horse. They have been found as far apart as North America and Mongolia.

Tyrannosaurus – 49 ft (15 m) of ferocious predator. Note that we have no idea of the actual skin colour.

Carnivores and Vegetarians

During the Cretaceous period, gigantic meat-eaters such as Tyrannosaurus, Daspletosaurus and Tarbosaurus ruled the land. The Tyrannosaurus Rex had up to sixty teeth that were as long as knives and just as sharp. Although the T-Rex was a fierce hunter, its huge size may have prevented it from moving quickly. It is possible that it charged at and head-butted its prey to stun them, then used its short arms to grip its victims while it ate them alive – though behaviour is difficult to judge from a fossil record alone. Much of the study of dinosaurs is based on supposition and guesswork – and until time travel becomes a reality, it always will be!

Compsognathus – meaning 'pretty jaw'.

Ornithomimids

The Velociraptor was made famous by the film *Jurassic Park* as a smaller version of Tyrannosaurus, hunting in packs. It may have used team work to single out and attack victims. Velociraptors were certainly well equipped to kill, with sharp claws, razor-sharp teeth and agile bodies.

Our experience of evolution and the modern world suggests that carnivore hunters are more intelligent than herbivores. In the modern world, for example, cows need very little intelligence to survive, while wolves and leopards are capable of far more complex behaviour. We apply the patterns we know to fill the gaps in the fossil record, but intelligence is one of those factors that are practically impossible to guess. If it were simply a matter of brain size, elephants would rule the land and whales would rule the sea.

Armour

One aspect of the age of dinosaurs that has practically vanished from ours is the use of armour for defence. It survives in tortoises, turtles and beetles, but otherwise, it has vanished as a suitable response to predators. By the end of the Mesozoic era, the arms race between predator and prey had produced some extraordinary examples of

Stegosaurus

armoured herbivores. The Stegosaurus, meaning 'covered' or 'roof lizard', is one of the best-known examples and evolved in the mid to late Jurassic period, some 170 million years ago.

Stegosaurus was a huge plant eater about the length of a modern 16 sixteen-wheel truck. The plates along its back would have made it much harder for a predator to damage a Stegosaurus spine. In addition, it had a viciously spiked tail to lash out at its enemies. Some dinosaurs, like the Ankylosaurus, even had their eyelids armour-plated.

Triceratops means 'three-horned face' and was named by Othniel C. Marsh, an American fossil hunter. It looked armoured for both attack and defence. It weighed up to ten tons and its neck protector was a sheet of solid bone – clearly designed to prevent a biting attack on that vulnerable area. It was very common 65 to 70 million years ago in the late Cretaceous period.

The camouflage dinosaurs used is unknown. Skin just doesn't survive the way bones do and, for all we know, some dinosaurs could have been feathered or even furred. Today's animals leave some clues, however. Living relatives of dinosaurs such as birds and crocodiles show how some dinosaurs may have been coloured. Large plant eaters like Iguanodon probably had green scaly skin and predators would have found them hard to spot among the forest ferns, very similar to today's lizards. Some carnivores may have also had green or brown colouring, to help them sneak up on prey. Successful hunters like the Velociraptor may have evolved light sandy skin if they hunted in desert regions or brown savannah, just as leopards have done today.

Like modern crocodiles, dinosaurs laid eggs. Some dinosaurs would look after these until they hatched, like the Maiasaur, which means 'mother lizard'. The evidence for this comes from the first one found in Montana, in a preserved nest containing regurgitated vegetable matter – suggesting that the parents returned to feed their babies as modern birds do. In addition, the leg bones of the fossilised babies do not seem strong enough to support the infants after birth, suggesting a vulnerable period spent in the nest. In comparison, modern-day crocodiles leave the egg as a fully functioning smaller version of the parent, able to swim and hunt.

In the skies of the Mesozoic, the reptile ancestors of birds ruled. There were many varieties, though most come under the species genus name of *Pterodactylus* – meaning 'winged fingers'. Of all species on earth, the link to birds from the Mesozoic era is most obviously visible, with scaled legs, hollow bones, wings and beaks. Many of them resembled modern bats, with the finger bones clearly visible in the wing. As might be expected, however, the Jurassic produced some enormous varieties. The biggest flying animal that ever lived may have weighed as much as a large human being. It was called Quetzalcoatlus – named after the feathered serpent god of Mexican legend. To support its weight it had a wing span of 39 ft (12 m) – like that of a light aircraft. It was almost certainly a glider, as muscles to flap wings of that size for any length of time would have been too heavy to get airborne.

Elasmosaurus

There were no icebergs in Mesozoic seas. In the strict sense of the word, there were no dinosaurs either, as dinosaurs were land animals. However, pre-historic oceans brimmed with a variety of strange and wonderful reptiles, like the giant sea serpent Elasmosaurus. The neck alone grew up to 23 ft (7 m) long and today people believe that 'Nessie', the Loch Ness monster, is a surviving descendant of an Elasmosaurus or some other plesiosaur, a similar breed.

Extinction

Hundreds of different dinosaurs roamed the earth seventy-five million years ago, yet ten million years later they had all but died out. Only the birds, their descendants, survived and what happened is still uncertain. An enormous crater in the Gulf of Mexico was almost certainly caused by a giant asteroid hitting earth. The impact occurred sixty-five million years ago, at the same time that the dinosaurs disappeared. Soil samples from the boundary between the Cretaceous and Tertiary periods – the moment of geological time known as the KT boundary – are found to be rich in iridium, an element commonly found in meteors and asteroids.

The asteroid would have hit earth at an incredible speed and dramatically changed the planet's atmosphere.

Huge clouds of rock and dust would have covered the sun, blocking out light and, crucially, warmth. Some animals lived through the changes; scorpions, turtles, birds and insects were just some of those resilient enough to survive. There is no definite explanation for why the dinosaurs vanished, although the asteroid strike is widely supported in the scientific community – at least for the moment

MAKING CRYSTALS

HAVING A CRYSTAL growing on your windowsill can be good fun. With food colouring, you can make them any colour you wish.

The problem is finding a suitable chemical. You may have seen copper sulphate and potassium permanganate in school. Both can be quite toxic and are therefore not easily available in local chemists. Your science teacher may allow you to have a sample, if you ask very politely.

For this chapter, we decided to use Potassium Aluminium Sulphate, better known as Alum powder. It is a non-toxic substance that used to be used to whiten bread. As with any household substance, you shouldn't get it in your eyes. It is available from the following website: www.panspantry.co.uk. 100 grams will cost you two pounds at the time of writing, not including postage and packing. That is enough for crystal making, but Alum can be used for fireproofing and tanning skin – as discussed in other chapters. It also works as an astringent on small cuts, or the crystals can be used as an underarm deodorant. You might want to get more. Alternatively, you can grow crystals with common salt or sugar.

You will need

- 10 grams of potassium aluminium sulphate (alum).
- A glass tumbler.
- A lolly stick (clean).
- Warm water.
- Thread.
- Small stones, preferably with sharp edges.

Method

1. Make sure the stones are clean – wash them thoroughly in running water.

2. Put enough warm water in the tumbler to cover the stones. (About a third of the cup.) Do not put the stones in yet.

3. Add the alum and stir furiously with the lolly stick until it stops dissolving easily. You may be left with a few grains at the bottom. Ignore them. You can either put the stones straight in or, for the classic look, tie a

thread around a small stone and the other end around the lolly stick, as in the picture. We did both.

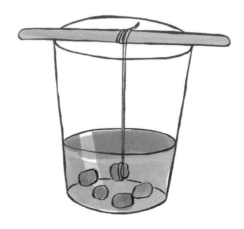

4. If you are intending to add food colouring, do it now. Show proudly to parents, who will pat your head for being a 'little genius'.

Evaporation is the key for these small crystals, so make sure it is in a warm place. It will take a few days for the first ones to appear, and the full effect can take a few weeks. Larger crystals can be made by repeating the process – after tying a small crystal to the thread.

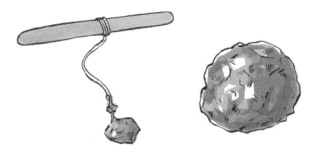

The crystal you see here is a picture of the one we grew – the one on the left, not the enormous thing. The huge circle came from the bottom of the glass and in many ways is more impressive than the actual crystal. It took about six weeks in total, and we refilled the alum once.

CHARTING THE UNIVERSE

THE ANCIENT GREEKS were the first recorded people to try and explain why natural events took place without reference to supernatural causes. Astronomy started to become a science and began its long journey from superstition to enlightened understanding. They were beginning to uncover the 'rules' of the universe but these often conflicted with the prevailing beliefs and the conflict between faith and science continues even today.

Thales was a Greek philosopher and explorer who lived in the 6th century BC. He travelled to Egypt to study geometry. On his return, he demonstrated a high level of mathematical skill, even predicting the eclipse of 585 BC. His legacy is the belief that natural events could have natural causes. It is true that he thought the world was flat and floated on water – but, on the other hand, he realised earthquakes could be explained as more than a bad-tempered Poseidon.

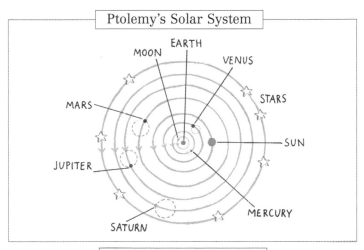

Ptolemy's Solar System

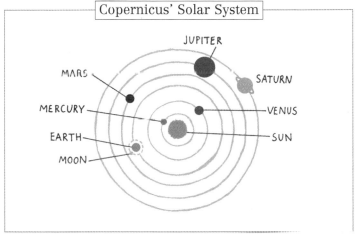

Copernicus' Solar System

Aristotle (384–322 BC) was one of the most influential of all Greek philosophers. He was a student of Plato, and became the teacher of Alexander the Great in Macedonia. He constructed three experimental proofs to show that the earth was round. He was the first to classify plants and animals. He thought that the earth was at the centre of the universe and that all the planets and stars were fixed in the heavens in a sphere around the earth. He believed earthquakes were caused by winds trapped beneath the earth.

Aristarchus flourished in the century after Aristotle and made a model to show that the sun was at the centre of things and not the earth. His theories were more scientific but history only briefly records his heliocentric ideas. However, no less a figure than Copernicus (see below) gives him credit in *De revolutionibus orbium coelestium*, writing, 'Philolaus believed in the mobility of the earth, and some even say that Aristarchus of Samos was of that opinion.'

Ptolemy of Alexandria was another gifted Greek astronomer. In AD 150, he published an encyclopedia (the *Almagest*) of ancient science with details and workings of the movements of the planets, showing an intricate mathematical system of circles within circles that buttressed his arguments for an earth-centred universe surrounded by unchanging spheres. This 'Ptolemaic system' was to rule the world of astronomy for 1,500 years.

Nicolaus Copernicus (1473–1543) was a Polish astronomer. Just before his death, he published his masterwork, *De revolutionibus orbium coelestium* – 'On the Revolution of the Celestial Spheres', which was to change humanity's view of the cosmos. Copernicus claimed that the sun was at the centre of the universe. This met with great hostility from the Christian Church, which had adopted the Ptolemaic geocentric (earth-centred) system.

Tycho Brahe (1546–1601) was a Danish astronomer who in 1572 saw a brilliant new star in Cassiopeia. This was a supernova – the explosion of a dying star, and in 1604 another supernova blazed forth in the sky. These events shattered a cornerstone of Ptolemaic thinking: that the outermost sphere was unchanging. The heavens had joined the renaissance.

Johannes Kepler (1571–1630) was Tycho Brahe's assistant and with his combined notes produced three laws of planetary motion. This enabled him to predict the positions of planets more effectively than Ptolemy.

Galileo Galilei (1564–1642) was an Italian scientist, who in 1609 took a telescope – then a new invention – and pointed it at the night sky. He discovered that the giant planet Jupiter had four moons clearly revolving around it in simple orbits, a miniature version of the Copernicus system. He published his discoveries, and in 1616 was warned by the Church to change his views. In 1632 he published *Dialogue Concerning the Two Great World*

Systems, ridiculing the Ptolemaic system. He was forced to recant and abandon his beliefs that the sun was at the centre of things and lived out his days under house arrest.

The Catholic Church later absolved Galileo from any wrongdoing. In 1989, a spacecraft was launched to study Jupiter and its moons; it was called *Galileo* and those moons are still collectively known as the Galilean moons.

These names should be known to all.

QUESTIONS ABOUT
THE WORLD – PART TWO

1. How do we measure the earth's circumference?
2. Why does a day have twenty-four hours?
3. How far away are the stars?
4. Why is the sky blue?
5. Why can't we see the other side of the moon?
6. What causes the tides?

1. HOW DO WE MEASURE THE EARTH'S CIRCUMFERENCE?

The simple answer is that we use Polaris, the Pole Star. Imagine someone standing at the Equator. From their point of view the Pole Star would be on the horizon – as in the diagram. If the same person stood at the North Pole, Polaris would be almost directly overhead. It should be clear, then, that in moving north, Polaris appears to rise in the sky. A sextant can confirm the changing angle.

The angle through which the Pole Star rises is equal to the change in the observer's latitude. If Polaris rises by ten degrees, you have travelled ten degrees of latitude.

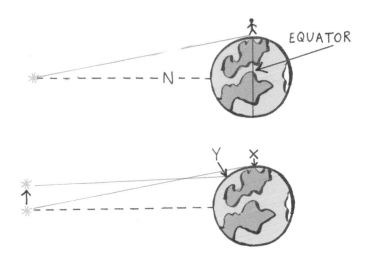

All the observer has to do is measure how far he has travelled when Polaris has risen by one degree. Multiply that distance by 360 and you have your circumference of the earth. Easy.

The actual circumference is 24,901 miles (40,074 km) around the Equator and 24,859 miles (40,0006 km) around the poles – or in rough terms 25,000 miles around, with a slightly fatter Equator. As you can see, this is not a perfect globe. The correct term is 'geoid', which just means 'shaped like the earth'. When you're a planet, you get your own word.

2. Why does a day have twenty-four hours?

Well, because we say it does. The modern world uses the Roman system of measuring time from midnight to midnight – as opposed to the Greek system of measuring from sunset to sunset. The Romans also divided daylight into twelve hours. This caused difficulties as summer hours would be longer than winter hours. When the system was made more accurate, it was sensible enough to double the twelve for the night hours. Most of the way we measure time is based on the number twelve, fractions and multiples of it, in fact – which is why we have sixty minutes and sixty seconds. The architects of the French Revolution were keen not only on introducing a decimal number system and metres to the world, but also a ten-day week, a hundred-minute hour and a hundred-second minute. Needless to say, no one else was quite as keen.

3. How far away are the stars?

Light travels at 186,000 miles (300,000 km) a second. In a year it would travel almost 6 million, million miles. (The American billion, or one thousand million, is now all-conquering in terms of usage. The British billion is traditionally one million million. Realistically, this is such a large number that it doesn't come up that often. There are no billionaires by the British standard, for example. A US 'trillion' is a thousand US billion. You may have noticed it's the same as the old British billion, but let's not overcomplicate this.)

Using the US definition, a light year is 6 trillion miles. That is a long way by anyone's standards.

The closest star to us is Proxima Centauri – about four and a third light years away. That is even further. To put it another way, the light from Proxima Centauri has taken four and a third years to get here. The actual star could have blown up yesterday, but we wouldn't know for almost five years.

The furthest stars we can see are more than a thousand light years away.

4. WHY IS THE SKY BLUE?

To understand this, it's important to understand that colour doesn't exist as some separate thing in the world. What we call blue paint just means paint that reflects light in certain wavelengths we have learned to call 'blue'. Colour-blind people have eyes that work perfectly well but are different from most other eyes in just this area – how they register light wavelengths. Take a moment and think about this. Colour does not exist – only reflected light exists. In a red light, blue paint will look black as there is no blue light to reflect. In a blue light, red paint will look black.

Now, the sky is blue because blue light comes in on a short wavelength and wallops into oxygen atoms of roughly the same size. When we look up and see a blue sky, we are seeing that interaction.

At sunset, we see more red because the sunlight is passing through many more miles of atmosphere at that low angle near the horizon. The blue light interacts with the oxygen and is scattered as before – but cannot reach the eye through the extra miles this time. Instead, we see the other end of the spectrum, the red light.

5. WHY CAN'T WE SEE THE OTHER SIDE OF THE MOON?

Until the late twentieth century, mankind had no idea what lurked on the dark side of the moon. This is because the same face was presented to observers on earth all the way through the lunar cycle.

The moon takes twenty-nine and a half days to go around the earth. It does actually rotate on its own axis, completing a full turn in ... twenty-nine and a half days. As these two are the same, it always shows the same face.

The best way to demonstrate this is with a tennis ball and a football. Mark the side of the tennis ball and place the football somewhere where it can't roll away – or have someone hold it. Now move the tennis ball around your earth, keeping the same side always inwards. By the time you have gone all the way around, the tennis ball will also have turned on its own axis.

6. What causes the tides?

Following neatly on from the last question, the answer is gravity – from the moon and the sun. The moon's massive presence overhead actually pulls oceans out of place. These two diagrams are deliberately exaggerated to show the effect. They are *not* to scale!

The seas move more easily than land, though the whole planet is actually affected. What happens in practice is that the earth's own spin produces two high and two low tides each day. It takes twelve hours to expose the other side of the earth to the moon's gravity, a little like squeezing a balloon twice around the middle in twenty-four hours. Both ends bulge to create high tides and then withdraw to create low tides.

Spring Tide – New Moon   Spring Tide – Full Moon

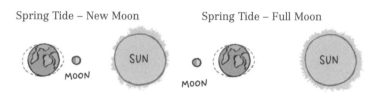

The diagram above is actually of a 'spring' tide, which occurs twice a month at the new and full moon. The name has nothing to do with the season. When the moon is in line with the sun and the earth, the tide is particularly strong. The weakest tides are known as 'neap' tides, and occur at the quarter moon, as in this diagram. The moon's effect is lessened by being out of line with the sun.

Neap Tide – Quarter Moon

FOSSILS

⟩━✦━⟨

Half a billion years ago there was no life on land and only worms, snails, sponges and primitive crabs in the seas. When these creatures died, their bodies sank into silt and mud and were slowly covered. Over millions of years, the sea bottom hardened into rock and the minerals of the bones were replaced, molecule by molecule, with rock-forming minerals such as iron and silica.

Eventually, this process turns the bones into rock – and they become known as fossils, a slowly created cast of an animal that died hundreds of millions of years ago. Other fossils are formed when dying animals fall into peat bogs or are covered in sand. As each new sedimentary layer takes millions of years to form, we can judge the age of the fossils from their depth. You can travel in time, in fact, if you have a spade. You can reach Roman times in just six or seven feet down in some places. To reach levels millions of years ago, you'll need to find a cliff where the layers are already revealed – like Lyme Regis and Charmouth in Dorset or the Lake District in Cumbria.

Those sea animals can move a long way in the time since they were swimming in dark oceans! Geological action can raise great plates of the earth so that what were undersea fossils can be found at the peak of a mountain or in a desert that was once a valley on the sea floor.

In parts of New Zealand, you can see the fossilised remains of ancient prehistoric forests in visible black bands on the seashore. This particular compressed material is coal and it burns extremely well as fuel. Oil too is a fossil. It is formed in pockets, under great pressure, from animals and plants that lived three hundred million years ago. It is without a doubt the most useful substance we have ever found – everything plastic comes from oil, as well as petrol for our planes and cars.

By studying fossilised plants and animals, we can take a glimpse at a world that has otherwise vanished. It is a narrow view and the information is nowhere near as complete as we would like, but our understanding improves with every new find.

Even the commonest fossils can be fascinating. Hold a piece of flint up to the light and see creatures that last crawled before man came out of the caves – before Nelson, before William the Conqueror, before Moses. It fires the imagination.

Here are some of the classic forms of fossils.

Ammonite. A shelled sea creature that died out at the KT boundary 65 million years ago (see Dinosaurs). Sizes vary enormously, but they can be attractively coloured.

Ammonite

FOSSILS

Trilobite. These are also a fairly common find, though the rock must usually be split to see them. Fossil hunters carry small hammers to tap away at samples of rock.

Sea urchin. Fossilised sea urchins and simple organisms like starfish are all very well, but remnants of woolly mammoths have been found in the south of England, as well as remnants of Jurassic period great carnivores and herbivores. However, you are likely to find a few ammonites on the Dorset coast in a single afternoon, while a Jurassic skeleton would be the find of a lifetime. That said, if you don't look, you won't find.

Trilobite

Sea urchin

ASTRONOMY – THE STUDY
OF THE HEAVENS

———※———

ASTRONOMY IS NOT ASTROLOGY. Astrology is nonsense. The idea that our lives can be affected by the flight of planets is not even slightly plausible. Venus may have been named after a goddess of love, but its movement can have no bearing on our own chances for romance. The planet could equally have been called by another name, after all. The first (and last) point about star watching is that it is science and not superstition – but the stories of ancient heroes like Orion can be fascinating. Knowing Orion chases Taurus works as a mnemonic – an aid to memory.

There are eighty-eight constellations that can be seen in the night sky at different times of the year and all the visible stars have names, or at least numbers. As the earth rotates, so their positions change and you can follow them through the seasons (see Star-Maps).

This chapter is an introduction to sky watching. Most of us live and work in noisy, artificial environments. Light pollution from cities hides the glories of the night sky, but those who are curious always find ways to explore beyond them. Naked-eye astronomy is easy and fun and can be done alone or with friends. This chapter will make you more familiar with the wonders of the universe.

Look at the stars! look, look up at the skies!
O look at all the fire-folk sitting in the air!
The bright boroughs, the circle-citadels there!

Gerard Manley Hopkins

Since the dawn of time, mankind has grouped stars into constellations, filling the heavens with heroes, gods and fantastic creatures. The myths and histories of lost civilisations can be found above us and help us understand the legends and stories that chart our own time.

One of the most easily recognisable constellations, and a great way to start finding your way around the skies, is **Ursa Major**, the **Great Bear**.

This constellation gets its name from the Greek legend of Callisto, a nymph transformed by Zeus into a she-bear. Many Native American tribes have also seen this constellation as a bear. Maybe the ancient Greeks sailed further than we realise! Particularly famous is the group of seven stars often called the **Big Dipper** or the **Plough**. In Cherokee legend, the handle of the Big Dipper is seen as a team of hunters chasing the bear who is visible high in the sky in spring until he sets on autumn evenings. Each

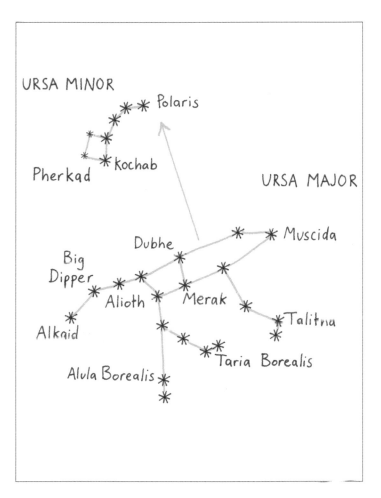

URSA MINOR

Polaris

Pherkad

Kochab

URSA MAJOR

Dubhe

Big
Dipper

Muscida

Alioth

Merak

Talitna

Alkaid

Taria Borealis

Alula Borealis

day they chase the bear further west. Boys, you will need your compass.

This distinctive star system has been noted by Shakespeare and Tennyson. In Hindu mythology, the Big Dipper is seen as the home of the seven great sages. The Chinese saw them as the masters of heavenly reality; the Egyptians, as the thigh of a bull. The Europeans saw a wagon and the Anglo-Saxons associated it with the legends surrounding King Arthur.

In ancient times, north could be plotted using the star **Alkaid**, in the Big Dipper. Today north can be found in **Ursa Minor**, a constellation that lies almost alongside Ursa Major. In Greek legend this constellation was named after Arcas, the son of Callisto. He too was changed into a bear and left to follow his mother eternally around the north celestial pole.

Finding north, and with it all other points on the compass, is as important as knowing your address. It is one of the first steps to understanding where you are. The key star is called **Polaris** (see previous page), the Pole Star for the northern hemisphere.

From the Big Dipper, mentally draw a line through the stars Dubhe and Merak, extend upwards five times its length and you hit Polaris. Face Polaris and you are facing north. If there is light pollution, it may be the only star visible in Ursa Minor.

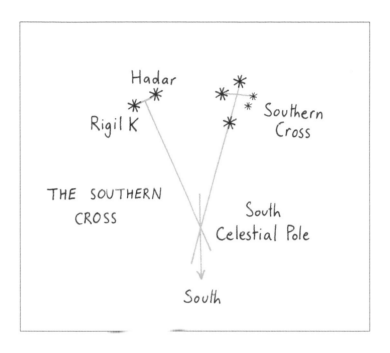

If you are in the southern hemisphere then finding south is just as important, and almost as easy. First identify the **Southern Cross** (see above) and mentally extend a line down from the long arm. To the left are two stars, Rigil Kentaurus and Hadar, known as the pointers. Extend a line down from between them until it crosses the first line. This point is directly above south.

On a clear night in winter in the northern hemisphere if you face south, away from the Pole Star, the constellation of **Orion** is the chief attraction. It is characterised by its three belt stars with the red star Betelgeuse above and Rigel below.

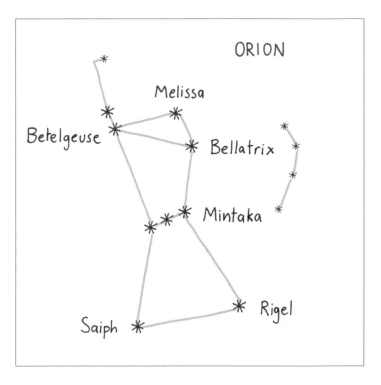

In Greek mythology, Orion was a great hunter. Artemis, the goddess of the moon and the hunt, fell in love with him and neglected her duties of lighting the night sky. As punishment, her brother, Apollo, tricked her into slaying him from afar with an arrow. When she realised what she had done, she put his body in the sky with his two war dogs, Canis Major and Canis Minor. According to ancient Greek astronomers, her grief explains the sad, cold look of the moon.

The brightest star in the sky is in Canis Major – **Sirius**, the **Dog Star**. Sirius rises in the east in late summer, at the heels of Orion, hunting with him through the winter.

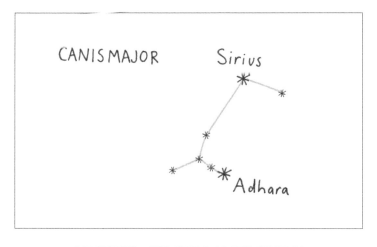

Above and to the right of Orion and his dogs is their prey, **Taurus**, the bull. Its red eye looks back nervously – the star Aldebaran. Since the time of the ancient Babylonians, some 5000 years ago, this constellation has been seen as a bull. Bulls have been worshipped since ancient times as symbols of strength and fertility. The Greeks saw the constellation as Zeus disguised as a bull. In this form he seduced the princess Europa and swam to Crete with her on his back. Only the forequarters are visible in the constellation, as it emerges from the waves.

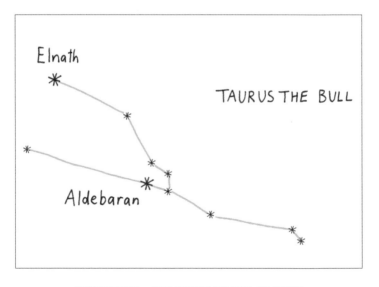

THE PLEIADES

Asterope
Taygeta
Maia
Celaeno
Alcyone
Atlas
Electra
Merope

In the shoulder of Taurus is the most famous open star cluster in the sky, the **Pleiades**, also known as the **Seven Sisters**.

The legend tells that the sisters were being chased by Orion and called out to Zeus to protect them. Zeus turned them into doves and placed them in the sky. In a Native American tale, the Pleiades are seven girls who are walking through the sky and get lost, never making it home. They remain in the sky, huddled together for warmth. The seventh sister is hard to see because she really wants to go home and her tears dim her lustre. On a reasonably clear night you should be able to pick out six of the sisters. The whole star cluster actually has more than 500 stars, but it is possible to see as many as nine with the naked eye.

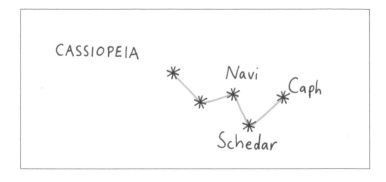

CASSIOPEIA

Navi
Caph
Schedar

On the other side of Polaris from the Big Dipper is the striking W-shaped figure of **Cassiopeia**. (Careful not to mix this up with the Little Dipper.) This is the most prominent constellation in the winter sky, visible all year round in the northern hemisphere. If the Big Dipper is low in the sky then the W of Cassiopeia will be high. It is not as accurate in finding north but it does point in the general direction of the Pole Star.

In Greek mythology, Cassiopeia was the Queen of Ethiopia. The Romans saw her as being chained to her throne and placed in the heavens to hang upside down, for boasting that her daughter, Andromeda, was more beautiful than Aphrodite. Arab cultures pictured the constellation as a kneeling camel.

Finding your way around the night sky can be quite a challenge for the beginner. In this chapter we have described a few of the brighter stars and constellations from which you will be able to explore further. There are many good periodicals about astronomy which will open up the sky to you. The stories that surround our heavens are wonderful and colourful and as easy as reading a road map, with a little work!

Remember that all stars twinkle – the light shifts and flickers as you concentrate on it. Planets do not. If you narrow your eyes, you can see the disc of Jupiter even without binoculars.

MAPS OF CHANGING BRITAIN –
ROMAN BRITAIN TO THE NORMAN CONQUEST

———※———

BRITAIN WAS SEMI-MYTHICAL to the Greeks and Phoenicians, who knew it as 'Cassiterides', the 'tin island' on the edge of the world. The Greek writer Herodotus wrote that he had heard of a sea on the other side of Europe from Greece, but personally doubted its existence. However, a Greek explorer, Pytheas, successfully circumnavigated Britain in the fourth century BC.

Map 1
Julius Caesar landed in 55 and 54 BC, but did not remain. In 43 AD, Rome invaded in force.

Map 2
When the Romans left in AD 440, Vortigern, the British king of Kent, invited mercenary Saxons to Britain to fight the Scots and Picts invading over Hadrian's Wall. The historian Gildas describes a dispute over payment and rations which led to a Saxon revolt. They ravaged the whole island as far as the western sea, achieved strongholds in East Anglia and Kent and so began the English conquest of Britain.

Map 3

5th to 8th century England went from a barbarous pagan land to a settled Christian one of eight Saxon kingdoms. This period is known as the Dark Ages, but the spread of Christianity also brought civilisation and the written word.

Map 4

Alfred became king in AD 871. This was a time when the 'great heathen army' of the Danes was fighting for control in England. Alfred defeated the Danes at Edington in Wiltshire. They signed the Treaty of Wedmore, which divided England, with Alfred controlling Wessex and the south and the Danes controlling northern England. Alfred created a powerful navy, built schools and ruled wisely. He is the only English king to be given the title 'Great'.

Map 5

Athelstan (reigned AD 924–939) was the first king of all England. He continued the re-conquest of what was left of the Danelaw area. In January 1066, Harold Godwinson was made king after the death of Edward the Confessor. He beat an invading Norwegian army at Stamford Bridge and marched south to face William of Normandy at Hastings in Sussex. William was victorious and was crowned king on Christmas Day, 1066.

Map 1

ROMAN
BRITAIN

Scale 1: 6,000,000
English miles
0   100

OCEANUS
HIBERNICUS

OCEANUS
GERMANICUS

MAEATAE

(Abercorn)   (Edinburgh)

Trimontium

Blatum Bulgium
(Birrens)

Hunnum
(Halton
Chesters)

Segedunum
(Wallsend)

HADRIAN'S WALL

Maglae Corstopitum
(Corbridge)

Luguvalium
(Carlisle)

Bremenium
(Kirby Thore)

Brocavum
(Brougham)

Veteris
(Brough)

Lavatrae

Vinovia   (Chester-le-Street)

(Ravenglass)

Cataractonium
(Catterick Bridge)

Monapia

Burium

PARISII

(Lancaster)

Olicana
(Ilkley)

Eburacum
(York)

Bremetenacum
(Ribchester)

Calcaria
(Tadcaster)

Mancunium
(Manchester)

Danum
(Doncaster)

(Wigan)

MELANDRA
CASTLE

Setea
Aest

Segontium
(Carnarvon)

Canovium

(Warrington)

Buxton

Lindum
(Lincoln)

Deva (Chester)

Sherwood
Forest

Motta

CORNAVII

Causennae
(Ancaster)

ORDOVICES

Vernemetum

CORITAVI

Branodunum
(Brancaster)

Viroconium
(Wroxeter)

Letocetum

Ratae
(Leicester)

Durobrivae

Venta Icenorum

Venonae

ICENI

Rockingham
Forest

SILURES

Magnae

Glevum
(Gloucester)

Forest
of Arden

CANOBANTES

DEMETAE

Gobannium
(Abergavenny)

Venta
Silurum

Camulodunum
(Colchester)

Octapitarum

Moridunum
(Carmarthen)

Isca
(Caerleon)

Corinium
(Cirencester)

Verulamium
(St. Albans)

DOBUNI

TRINOVANTES

Londinum

Sab(rina Aest (Severn)

Aquae Sulis
(Bath)

Cunetio

Durobrivae

Regulbium
(Reculver)

Calleva Atrebatum
(Silchester)

ATREBATES

Durovernum
(Canterbury)

Rutupiae
(Richborough)

Herculis Pr

BELGAE

Sorbiodunum
(Salisbury)

CANTII

Portus

ANDERIDA SILVA

Dubrae
(Dover)

Ilchester

DUROTRIGES

Venta Belgarum
(Winchester)

Anderida

Isca Dumnoniorum
(Exeter)

Durnovaria
(Dorchester)

Clausentum

REGNI

Regnum

DAMNONII

Vectis

LITTUS SAXONICUM

Bolerium Pr.

Ocrinum Pr.

OCEANUS BRITANNICUS

MAPS OF CHANGING BRITAIN – ROMAN BRITAIN TO THE NORMAN CONQUEST

Map 2

THE ENGLISH CONQUEST OF BRITAIN

Scale 1: 6,000,000
English miles
0   100

Approximate extent of English Conquest ...	... in 550	... in 600
	... in 650	... in 550

MAPS OF CHANGING BRITAIN – ROMAN BRITAIN TO THE NORMAN CONQUEST

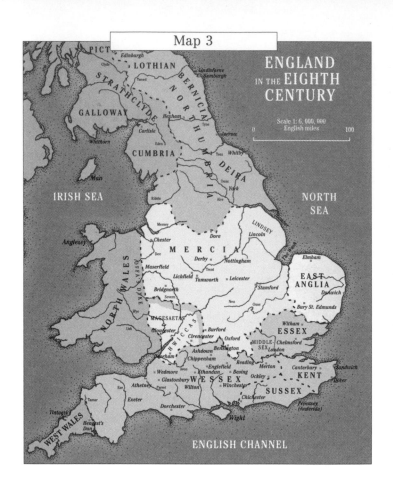

Map 3

ENGLAND
IN THE EIGHTH CENTURY

Scale 1: 6,000,000
English miles
0   100

PICTS
Edinburgh
Clyde
LOTHIAN
Lindisfarne
Bamburgh
STRATHCLYDE
Tweed
BERNICIA
NORTHUMBRIA
GALLOWAY
Hexham
Whithorn
Carlisle
Eden
Tyne
Jarrow
CUMBRIA
Tees
Whitby
DEIRA
Man
Ribble
Ouse
York
Aire

IRISH SEA

NORTH SEA

Mersey
Dore
LINDSEY
Lincoln
Anglesey
Chester
Dee
MERCIA
Derby
Nottingham
NORTH WALES
OFFA'S DYKE
Maserfield
Trent
Elmham
Lichfield
Tamworth
Leicester
EAST ANGLIA
Bridgnorth
Severn
Stamford
Dunwich
Wye
Nen
Ouse
MAGESAETAS
Bury St. Edmunds
Lugg
Gloucester
Burford
HWICCAS
Cirencester
Oxford
Witham
ESSEX
Avon
Ashdown
Benington
MIDDLE-SEX
Chelmsford
Porlock
Chippenham
Reading
London
Wedmore
Eng
Merton
Ethandun
Basing
Ockley
Canterbury
KENT
Athelney
Glastonbury
WESSEX
Winchester
Sandwich
Exe
Parret
Wilton
SUSSEX
Dover
Tamar
Exeter
Chichester
Pevensey
(Anderida)
Tintagel
Dorchester
Hengest's
Dun
Wight
WEST WALES

ENGLISH CHANNEL

MAPS OF CHANGING BRITAIN – ROMAN BRITAIN TO THE NORMAN CONQUEST

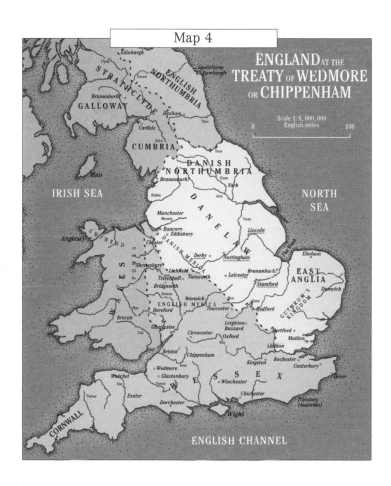

Map 4

ENGLAND AT THE TREATY OF WEDMORE OR CHIPPENHAM

Scale 1: 6,000,000
English miles
0   100

Map 5

ENGLAND
ON THE EVE OF THE
NORMAN
CONQUEST

Scale 1: 6,000,000
English miles
0   100

Labels on map:

Edinburgh
Clyde
LOTHIAN
Carham
Holy Island
Bamburgh
BERNICIA
SUDREYS
SCOTLAND
Tyne
CUMBRIA
Chester
-le-Street
Durham
Jarrow
WESTMORELAND
NORTHUMBRIA
(MORCAR)
Eden
Tees
Man
IRISH SEA
Lancaster
Derwent
York
Stamford Bridge
Fulford
NORTH
SEA
Ribble
Ricall
Aire
Anglesey
Manchester
Mersey
Gainsborough
Lincoln
MERCIA
(EDWIN)
Chester
Dee
The Wash
Norwich
NORTH WALES
Shrewsbury
Stafford
Derby
Nottingham
Trent
Leicester
Stamford
Ely
EAST
ANGLIA
Thetford
Ringmere
Leominster
Severn
Coventry
Warwick
Worcester
Northampton
WALTHEOF
Nene
Huntingdon
Bedford
Ipswich
Teifi
Wye
Hereford
GYRTH
Towy
Usk
Gloucester
Oxford
Ouse
Hertford
Maldon
Sherston
Fyfield
Wallingford
Waltham
London
LEE
Thanet
Bristol
Brentford
Canterbury
Sandwich
Wedmore
Wells
Glastonbury
WESSEX
(HAROLD)
Otford
Dover
Watchet
Parret
Penselwood
Winchester
Battle
PWINE
Exe
Tamar
Exeter
Southampton
Bosham
Hastings
Pevensey
(Anderida)
Wight
ENGLISH CHANNEL

CLOUD FORMATIONS

IT REALLY IS AMAZING just how many times you can look up at the sky in a lifetime and say 'I can never remember, is that Cumulocirrus, or Strato-whatsit?' Everyone is taught them at school and, frankly, we all forget. You'll read them now and when you *really* want to know, you'll have forgotten. The solution is to get spare copies of the book, so that you always have one with you.

THERE ARE ONLY THREE BASIC TYPES OF CLOUDS

Cirrus

The previous image is of **Cirrus** – light, wispy clouds, which can be as high as fifteen thousand feet and are made of ice crystals. The formation is sometimes referred to as 'mare's tails'.

After that comes the most common – **Cumulus**. These are the fluffy cotton-wool clouds you can see on most days.

The last member of our big three is **Stratus** – a dark, solid blanket of cloud at low level.

Cumulus

Stratus

All cloud formations are combinations of these three basic forms. The only other word that crops up is Nimbus – meaning a dark grey rain cloud. You could for example, see Cumulonimbus, which would be large and fluffy, but dark and just about to rain. The leading edge of a storm

THE MAIN CLOUD FORMATIONS

HIGH ALTITUDE
(above 18,000 ft/5500 m)

Cirrus – high and wispy

Cirrostratus – high thick layer

Cirrocumulus – high cotton wool

Cumulonimbus – cotton-wool storm clouds

MEDIUM ALTITUDE
(6,500–18,000 ft/2000–5500 m)

Altostratus – medium-height heavy band

Altocumulus – medium-height cotton wool

LOW ALTITUDE
(up to 6,500 ft/2000 m)

Stratus – heavy flat layer

Stratocumulus – fluffy and flat combined

Cumulus – cotton wool

Nimbostratus – raining flat layer

is usually Cumulonimbus. Nimbostratus would be a heavy dark layer covering the sky and again just about to pour down.

You know a storm is coming when you see Stratus and Stratocumulus cloud formations getting lower. If the clouds descend quickly into Nimbostratus, it is time to find shelter as the rain will be coming at any moment. If you happen to have a barometer, check the mercury level. A sudden drop in pressure indicates a storm is on the way.

These ten can be further subdivided, with names such as Cumulonimbus Incus, an anvil-shaped storm cloud often called a 'thunderhead'. For most of us, however, just remembering and identifying all ten major types would be enough.

MAKING CLOTH FIREPROOF

PERHAPS THE MOST impressive use for alum (potassium aluminium sulphate) is in fireproofing material. This could be very useful for tablecloths where there is a fire hazard, as in a laboratory or on a stage. It works with any porous cloth, but should not be considered foolproof. To demonstrate it, we used household dusters.

First prepare a solution of alum and water. Hot water works best in dissolving the powder. 1 lb 1 oz (500 g) of alum dissolves easily in a pint of water. Dip the material you wish to fireproof in the solution and make sure it is completely covered. Remove immediately and leave to dry. Be careful not to let it drip onto valuable carpets. If you leave it outside and it happens to rain, it will probably still work.

Once dry, the cloth should be a little stiffer than usual, but otherwise unchanged. An untreated duster burned almost completely in twenty seconds. The treated duster could not be lit, though there was some light charring after thirty seconds of applied flame.

PINHOLE PROJECTOR

You will need

- Long cardboard tube.
- Sheet of card or A4 paper.
- White tissue paper.
- Adhesive tape.

THE SUN IS A BRIGHT shining star that gives life to our planet. It is also very dangerous to look at directly and can damage your eyes. How then do we study the eclipses and transit of planets across its surface? With a pinhole projector, one of the best tools to view solar events as they take place.

As the name implies, a pinhole projector projects an image (upside down) of the sun so you can watch an eclipse without actually looking at it.

To build the projector, use glue or adhesive tape to attach the sheet of card at one end of the tube, as shown in the diagram. Make a small hole in the card over the tube. Attach a thin sheet of tissue paper to the other end with glue.

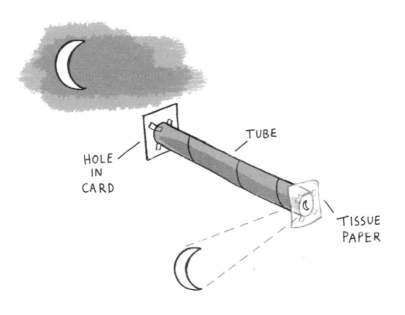

To make it work, simply lift the card end towards the sun and an image will be projected onto the tissue paper 'screen' at the other end. Remember, only look at the projected image.

PINHOLE PROJECTOR

You can also build a projector with a magnified image, which is far more impressive and can be focused. You will need a telescope or binoculars and two pieces of white card.

First cut a hole in one of the pieces of card so that it fits over the end of your telescope or one side of your binoculars to shield the image from unwanted light. Now aim the telescope (or binoculars) at the sun and hold the other piece of white card a couple of feet from the eyepiece. An image of the sun should appear, which you can sharpen by changing the focus or moving the card. Never, *ever* look through a telescope while focusing or aiming it at the sun.

With a similar contraption, we saw the transit of Venus cross the sun in 2004. These transits usually come in pairs and the next is due on 6 June 2012. Partial eclipses are not particularly rare, though whether you can see one depends on where you are in the world. Here is a list of the more impressive total solar eclipses coming up in the next few years. Each date is followed by the latitude and longitude that will give the best view. For Britain (with a longitude of zero at Greenwich), the one in 2015 will be the easiest to see – especially from the western side of Scotland.

1. 01/08/2008	Latitude: 65.6N	Longitude: 72.3E
2. 22/07/2009	Latitude: 24.2N	Longitude: 144.1E
3. 11/07/2010	Latitude: 19.8S	Longitude: 121.9W
4. 13/11/2012	Latitude: 39.9S	Longitude: 161.3W
5. 20/03/2015	Latitude: 64.4N	Longitude: 6.6W
6. 09/03/2016	Latitude: 10.1N	Longitude: 148.8E
7. 21/08/2017	Latitude: 37.0N	Longitude: 87.6W
8. 02/07/2019	Latitude: 17.4S	Longitude: 109.0W
9. 14/12/2020	Latitude: 40.3S	Longitude: 67.9W

In 2038, there will be seven solar and lunar eclipses. A total lunar eclipse is a strange and wonderful sight, as light is scattered by the earth's atmosphere to turn the moon a dark red, as if it were made of copper.

Set up your pinhole projector and enjoy the sights.

THE COMMONWEALTH

FACTS AND FIGURES

The Commonwealth is an organisation of fifty-three nations. With the exception of Mozambique, which joined at the end of the twentieth century, the other fifty-two were all part of the British Empire. In fact, the Commonwealth was created as the peaceful twilight organisation of that empire. It has been largely successful, remaining a surprisingly influential group today. More than 1.8 billion people live in Commonwealth countries and Queen Elizabeth II broadcasts an address to them on Commonwealth Day (the second Monday in March) each year. All fifty-three take part in the Commonwealth Games.

The * symbol indicates the constitutional monarchies where Queen Elizabeth II is head of state. Up to 1947, there were no republics in the British Commonwealth. To allow India to remain a member whilst becoming an independent

1. Antigua and Barbuda*
2. Australia*
3. The Bahamas*
4. Bangladesh
5. Barbados*
6. Belize*
7. Botswana
8. Brunei Darussalam
9. Cameroon
10. Canada*
11. Cyprus
12. Dominica
13. Fiji Islands.
14. The Gambia
15. Ghana
16. Grenada*
17. Guyana
18. India
19. Jamaica*
20. Kenya
21. Kiribati
22. Lesotho
23. Malawi
24. Malaysia
25. Maldives
26. Malta
27. Mauritius
28. Mozambique
29. Namibia
30. Nauru
31. New Zealand*
32. Nigeria
33. Pakistan
34. Papua New Guinea*
35. St Kitts and Nevis*
36. St Lucia*
37. St Vincent and the Grenadines*
38. Samoa
39. Seychelles
40. Sierra Leone
41. Singapore
42. Solomon Islands*
43. South Africa
44. Sri Lanka
45. Swaziland
46. Tanzania
47. Tonga
48. Trinidad and Tobago
49. Tuvalu*
50. Uganda
51. United Kingdom*
52. Vanuatu
53. Zambia

republic, the word 'British' was dropped from the description. Today, thirty-two members are republics and five (Brunei, Lesotho, Malaysia, Swaziland and Tonga) have national monarchs of their own. All, however, accept the British monarch as 'Head of the Commonwealth'. In addition, Crown Dependencies like Guernsey, Jersey and the Isle of Man can take part in the Commonwealth Games.

This is an extract from a speech on the Commonwealth by the Right Honourable Owen Arthur, Prime Minister of Barbados:

> It is the oldest living political association of states, yet in many ways the most adaptable to modern realities and thus the most responsive to the changing needs of its membership. It is rich in its diversity, yet remarkable in its cohesiveness, forged in no small measure by the sense of common identity we derive from our shared historical experience and the administrative, legal and institutional structures.

The Commonwealth promotes democracy, equality and good governance not only amongst its own members, but also throughout the world. It gives aid and disaster relief to its members and provides funds for development and the eradication of poverty.

NAVIGATION

———✦———

THE FIRST THING TO understand is that a compass points north because it is magnetic and the earth has a magnetic field caused by the rotation of a liquid metal core. The magnetic north pole happens to correspond reasonably well with the true pole – but they are not the same. Magnetic south is off Antarctica and can be sailed over. Magnetic north is near the Canada/ Alaska border. They are both very deep within the core of the planet and move over time.

If you are interested, a compass will actually jam on the magnetic poles as it tries to point either 'up' or 'down' – ninety degrees to the surface. A gyroscopic compass is invaluable in such circumstances – that is, a gyro that has been set to point north and then holds its position regardless of changes in direction. Pilots find gyroscopic compasses invaluable. The International Space Station (ISS) has thirteen of them.

Admiralty charts plot the lines of 'magnetic variation' across the globe, showing whether the variation from true north is to the east or west and increasing or decreasing. As you can imagine, this is crucial for navigation. A compass in New York will be approximately 14° W off true north. If you were plotting a course north, you would have to subtract 14 degrees from your compass direction. If the difference was 14° E, 14 degrees would have to be added.

The compass is the universal means of finding your position anywhere on the surface of the planet. The earth rotates east, so in *both* hemispheres, the sun rises in the east and sets in the west. It is true, however, that water swirls the other way down plugholes and toilets in the southern hemisphere.

The figure opposite shows the thirty-two points of the compass. In the northern hemisphere when the sun is at its highest point in the sky, it will be due south. In the southern hemisphere this noonday point will be due north.

KEY: Read the word 'by' for the symbol –, so N–NE is north *by* north east.

The hemisphere can be indicated by the movement of the shadow cast by the sun: clockwise in the north and anticlockwise in the south. This shadow can also be a guide to direction.

Shadow Stick

One hour before noon, place a three-foot stick upright on flat ground and mark where the tip of the shadow falls - point 'a'. At one hour past noon, mark where the tip of the new shadow falls - point 'b'. Draw a line from 'a' to 'b' and you have an east-west line, 'a' being west. This will only work when you take noon as your centre point. When you have your east-west line, dissect it at right angles and you have a north-south line. With 'a' on your right and 'b' on your left, you are facing south. This works in both hemispheres. Feel free to heat your brains up trying to explain why.

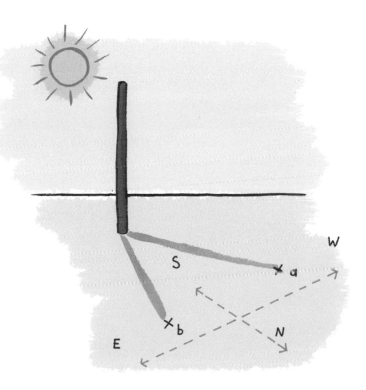

FINDING DIRECTION WITH A WATCH

A watch with two hands can tell the direction. It must have the correct local time (excluding daylight saving: this is when you put the clocks back and forward – 'spring forward and fall back' – so in summer you should *add* an hour to use this technique). The nearer to the Equator you are, the less accurate this is.

In the northern hemisphere, hold the watch horizontally. If it's summer, wind it back an hour; if it's winter, wind it on an hour. Point the hour hand at the sun. Bisect the angle between hour hand and 12 to give you a north–south line. In the southern hemisphere point 12 at the sun, and the mid point between 12 and the hour hand will give a north–south line.

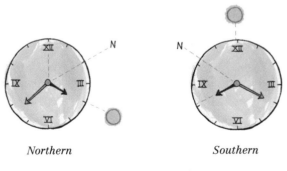

Northern   *Southern*

Needle Compass

Get a piece of ferrous (meaning iron) wire – a sewing needle is ideal – and stroke it in one direction repeatedly against silk. This will magnetise it. Suspend the needle on a length of thread and it will point north.

Stroking the wire with a magnet in one direction will work better than silk. This aligns the atoms in the needle. Heating the needle also works, though not as reliably. Try it and see.

If you have no thread then you can also float the magnetised needle on a piece of tissue paper or bark on the surface of water and it will turn to indicate north.

An old-style razor blade can also be used as a compass needle. Rub it against the palm of your hand (carefully!) to magnetise it, then suspend it to get the north–south line.

Use as many methods as you can to get your bearings, then mark out your compass, check all your readings against the sun and keep your needle magnetised.

DIRECTION BY THE HEAVENS

To find north in the night sky you need to find Polaris, the Pole Star. This is discussed in the Astronomy chapter. There are other indicators in the night sky which can be used. The rising of the moon can give a rough east–west reference. If the moon rises before the sun has set, the illuminated side will be on the west. If it rises after midnight, the illuminated side will be on the east.

Stars themselves can also be used to indicate direction. If you cannot find Polaris or the Southern Cross, get two sticks, one shorter than the other. Stick them in the earth and sight along them as shown to any star except the Pole Star. From the star's apparent movement, you can work out the direction you are facing!

If the star you are lined up on appears to be rising, you are facing east. If it appears to be setting (or falling), you are facing west. If the star seems to move right then you are facing south and if it moves left, you are facing north. These are only approximate directions and will be reversed in the southern hemisphere.

Being able to find your bearings at any time of day and night is a pretty impressive thing to know, but try not to show off your knowledge. Keep it safe for a time when you may really need it. As the Scouts say, 'Be prepared'.

LIGHT

W**HAT IS LIGHT?** Without the human sense of sight, the word 'light' would have no meaning. Light enters our eyes and we 'see' things. Seeing things is a mental sensation and light is the physical cause of this. The mental effect that light causes is still one of the mysteries of the mind but we do understand a great deal about light on the physical side.

The thing we see might have its own light source – like a light bulb, or light might be reflected off it from somewhere else – like the sun. We see most things by this reflected or borrowed light.

The origin of any light source will begin with the vibration of atoms. A light bulb, for example, uses electricity to heat a filament to the point where it gives out energy in the form of white light. That light travels at about 186,000 miles per second (300,000 km/s) in empty space. It travels in waves or a steady flow of waves, like ripples on a pond. The waves have a very short wavelength (this is measured from the crest of one wavelength to another): 1/40,000 inch (0.00006350 cm) to 1/80,000 inch (0.00003175 cm), depending on the colour.

When light shines on a non-luminous body (like a table),

it stimulates the atoms to varying degrees. Some atoms absorb all the light that falls upon them, while other atoms absorb some of the light but allow the rest to be reflected. The light finally reaches the eye, producing on the retina an image of the object viewed. Thus we 'see' and recognise the different parts of an object.

Light waves of various wavelengths create the sensation of colour when they fall on the eye. These waves can be identified by passing light through a prism, a coloured strip called 'the spectrum' being produced, red at one end and passing through orange, yellow, green, blue and indigo to violet at the other end. 'Richard Of York Gave Battle In Vain' is a good way to remember the colours of the spectrum.

By mixing colours, using coloured glass and a white background, any colour whatever can be produced, including some that are not present in the spectrum, like brown. The eye cannot tell the composition of the light that produces any given colour; for example, the colour yellow is a simple colour, but may be produced by mixing red and green in the correct proportions. Whiteness is caused by a mixture of all the simple colours. The classic way to see this is to colour a card disc with the shades of the rainbow and punch a pencil through the middle. When spun quickly, the colours will blur into whiteness.

When we look at a raindrop, we call it transparent, and think that the light goes straight through. Actually some of it is reflected from the inner surfaces. The light is bent or 'refracted' as it enters the raindrop and again when it leaves.

Raindrops act in the same way as rough prisms of glass or ice and cause rainbows. The drops of water split up the sunlight into the colours of the rainbow by 'refracting' each of the different colours of light to a different degree.

You will always find that your shadow points directly to the middle of the rainbow. You might also hear of a 'crock of gold' where the rainbow ends. Unfortunately, a rainbow has no end. As you move your position so the rainbow will move with you. Curiously no two people will ever see exactly the same rainbow. They generally appear when the sun is fairly low in the early morning and afternoon. The lower the sun, the higher the bow.

Colour is an affair of the mind, whilst light is purely physical, but you cannot have one without the other.

LIGHT

COMMON BRITISH TREES

In ABOUT 4000 BC the whole of Britain was covered by woodland. The natural environment in temperate regions does favour trees, but human activity does not. Most of Britain has been cleared of its ancient woodland by the migration of people using fire and axe. Millennia later, Britain rose as the pre-eminent sea power on the back of English oak.

To our ancestors, forests were an essential part of the rural economy, providing timber for houses, animals to trap, charcoal for fuel, wild mushrooms and herbs. A system evolved called 'coppicing', where an area of undergrowth and small trees was grown for periodic cutting, managed like any other crop.

There are still hundreds of different varieties of trees all over the country, too many to list here. However, it is a good idea to know the common trees of towns and countryside and how to identify them. It is as important to understand the earth around you as it is the heavens above. It might even come in useful when it comes to making things from wood.

The identification of trees can seem a formidable task. Overhearing a botanist wandering around murmuring

'*Sequoiadendron giganteum*', may be a little scary. There is nothing wrong with recognising varieties by a process of categorisation and elimination, however. Here are a few of the most common – the ones we think *everyone* should know.

THE ENGLISH OAK (*QUERCUS ROBUR*)

The most important forest tree in north-west Europe, varying in size from 70 to 130 ft (21–40 m). Oak forms extensive woods and is fantastically long-lived, often surviving for centuries. Beech is more common in the south of England and ash and pine in the north. However, in the English midlands, oak reigns supreme.

- **Bark:** Grey, with fine cracks and ridges. The twigs bear rounded but pointed buds.
- **Leaves:** Clear and separate lobes.
- **Fruit:** The familiar acorn, on long stalks in clusters. Groups of seedlings can often be seen because squirrels bury them to ensure a winter food supply.

English Oak and acorn

THE COMMON LIME (*TILIA X EUROPAEA*)

A magnificent tree, tall and elegant, (75 ft/23 m), a well-established hybrid of the small- and large-leafed limes which form woods on the limestone slopes around Bristol and south Yorkshire. Traditionally planted in avenues. Lime flowers when dried can make a wonderful tea that is helpful in aiding sleep.

Common Lime leaf

- **Bark:** Grey or grey-brown, long ridges and cracks and lumps.
- **Leaves:** Heart shaped, narrowing to a point, dark green, with tufts of white hair underneath. The twigs themselves are hairless.
- **Fruit:** rounded Rounded and faintly ribbed when ripe; fall in clusters.

Common Lime

HAWTHORN (*CRATAEGUS MONOGYNA*)

Hawthorn

One of the most familiar sights in lowland Europe as well as Britain. A dense thorny shrub, but which can also grow as a small tree (40 ft/ 12 m). Common on all soils as a hedge plant.

One of the old names for the Hawthorn blossoms is 'May', which led to the ancient advice: 'Ne'er cast a clout til May is out' – don't remove a vest until spring comes. It's still good advice.

- **Bark:** Light brown and flaky.
- **Leaves:** Variable but lobed.
- **Fruit:** Distinctive small, deep red, round haws (fruit).
- **Flowers:** Appear in early June in white clusters, turning pink as they mature.

Hawthorn haws and leaf

Silver Birch (*Betula Pendula*)

Silver birch is a common tree of dry and sandy soils. Fast-growing, they rarely last longer than 100 years. At maturity, they can reach 100 ft (30.5 m). Distinguished by stiff branches and dropping twigs. Spring sap can be tapped into bottles and tastes like clear, sugary water. However, if the hole is left unplugged or too much is taken, the tree will die.

Silver Birch leaves

- **Bark:** Pinkish brown in young trees, turning white with black patches.
- **Leaves:** Oval with a long pointed tip, serrated around the edges.
- **Fruit:** Catkins (small twig). Shed lots of papery winged seeds in autumn, yellowy in colour.

Silver Birch

COMMON BRITISH TREES

133

Beech (*Fagus Sylvatica*)

Beech

An impressive tree, which can grow up to 140 ft (42.5 m), with an enormous spreading crown. Dominant in chalky soils. Beeches survive for centuries, growing immense, twisting trunks. The wood is extremely hard and used in school carpentry benches.

- **Bark:** The trunk is smooth and grey, branching out horizontally.
- **Leaves:** Oval and pointy with clear veins at the edges. Spring – yellowy, summer – dark green, autumn – a rich brown. Twigs are brown with narrow pointed buds.
- **Fruit:** A hard glossy brown nut in a hairy shell. The inner nut can be eaten and tastes delicious.

Beech leaf and nut

Horse Chestnut (*Aesculus Hippocastanum*)

A very familiar suburban tree, but not a native of Britain. Introduced from south-east Europe in the seventeenth century. Shakespeare never played conkers! The name comes from the resemblance to the edible chestnut. Can grow up to 80 ft (24 m).

- **Bark:** Orangey-brown.
- **Leaves:** Green oval, with an unmistakable spread.
- **Fruit:** Dark, sticky buds, beginning to open in April, forming green, spiny fruit cases towards the end of summer. These fall in September, the shells splitting to reveal brown nuts known as conkers.

Conkers and Horse Chestnut leaf

Horse Chestnut

Sycamore (*Acer Pseudoplatanus*)

Sycamore

By far the most common maple in Britain. A beautifully domed tree that can grow up to 120 ft (36.5 m). In Britain, it is the most common tree in upland areas.

- **Bark:** Smooth and grey, becomes cracked and rough. Twigs are short and have fat green buds.
- **Leaves:** Large and green with five pointed lobes.
- **Fruit:** Form in pairs and spiral when in flight.

Sycamore leaves

COMMON BRITISH TREES

ASH (*FRAXINUS EXCELSIOR*)

The trunk is often long and straight. In northern England it is the main hedgerow tree, the only British tree that is of the olive family. Usually 70–80 ft (21–24.5 m), but can grow up to 140 ft (42.5 m).

- **Bark:** Smooth and grey, long cracks with age. The twigs are sticky, and have big black buds.
- **Leaves:** Pinnate (meaning 'pairs either side of stem'). Green, small and pointy.
- **Fruit:** In October, it sheds seeds that resemble a long, narrow, brown wing.

Ash leaves and fruit

Ash

COMMON BRITISH TREES

141

STAR MAPS: WHAT YOU SEE WHEN YOU LOOK UP...

Facing south in the northern hemisphere, turn the book so the current month of the year is at the bottom. This will be accurate at around 11 in the evening.

Facing north in the southern hemisphere, turn the page again to put the correct month at the bottom and these are the constellations visible on a good clear night at 11 p.m.

STAR MAPS: WHAT YOU SEE WHEN YOU LOOK UP

HUNTING AND COOKING
A RABBIT

———※———

U NDER UK LAW, it is illegal to use an airgun under the age of fourteen. From fourteen to seventeen, an air rifle may be given as a gift and used legally, though it is still illegal to buy one before seventeen. It is also worth considering the lesser-known fact that if an air rifle is used in the commission of a crime of any kind, it is considered to be 'a firearm' – with serious penalties applied.

The legal limit for an air rifle sold without a firearm certificate is twelve foot-pounds of pressure. There are two main types: those you cock by pulling the barrel back on itself and those that work from compressed air held in a canister under the barrel. The type that cocks is cheaper and doesn't need recharging every 100 shots. It should last practically for ever. Target shooting can be a highly enjoyable pursuit, but a powerful air rifle can also be used to hunt game – rabbits, pheasants and pigeons.

To do so is not a game, nor is it a sport. We believe the experience is valuable as it gives an insight into the origin of those neat meat packages you see in supermarkets. The aim, however, should be to get lunch – if you kill something, you have to eat it.

It is possible to hunt rabbits with a bow and arrow, but the movement involved in pulling the string back tends to spook them and we cannot recommend this unless you are capable of holding a drawn bow motionless for ten or twenty minutes. Believe us when we say it is extremely hard to hunt rabbits with a bow. You tend to lose the arrows as well.

Before you go anywhere near a live shoot, spend time with a target set up at twenty or thirty yards. You can make a simple bull's-eye by drawing round two cups in circles on a bit of paper. Bring drawing pins with you to fix it to a tree.

A yard is a normal walking pace, so it's easy to set up the range. You need to be certain that when you have something in your cross-hairs, the pellet will hit where you point it. The method here is to find a steady aiming spot, a tree stump, for example, and fire five shots at the bull, taking note of where they hit. If you are steady, they should be close together. If all of your shots at bull are hitting low and to the left, say, you'll need to adjust your sights to fire up and right. Practise until you can hit the bull regularly. You should not stint when buying the pellets – you want ones that are checked for quality and heavier than usual. Don't bother with the pointed-head pellets. Weight is far more important. It does cut the range a little, but is more likely to result in a clean kill.

FINDING THE RABBIT

Get out into the countryside, for a start. It is not illegal to fire an air rifle within the confines of your own garden, unless the pellets pass outside the boundary, in which case you are likely to have an armed police team turn up. Be sensible – look for rabbits where there are fields. Note that it is also illegal to walk around with an uncovered weapon, but if the weapon is in a carrying sleeve, you can walk on public land with one. That said, the law can change and may have by the time you read this. Check with your local police station before taking an air rifle on a public highway.

Rabbits never move far from their warrens. If you have ever seen one in a field, their burrow will be very, very close by. The best thing to do will be to note where they are seen over a period of time, to have an idea of where to find them. It is possible to come upon them on a ramble, but it's a little hit and miss.

This is one place where exercising a little common sense wouldn't hurt. Go and ask the owner of the land if you can shoot rabbits. If it's a field, or a farm, there's a very good chance they will say yes. Rabbits breed like maniacs and are not much loved by farmers. You may even be given directions to the best spots. However, shooting a pheasant will provoke a very angry reaction. These birds are a cultivated crop whose value lies in the shooting fees paid for them. Poachers are not appreciated and illegal poaching carries heavy penalties.

Once in the area, find the warren. You might see rabbits in the distance, but as you come closer, they will all vanish. After you locate the complex of burrows, you should get between 60 and 90 feet (18–28 m) away (20-30 yards) – the effective range. Much further and you are likely to miss a kill. Much closer and they will remain nervous in your presence.

Have a pellet ready in the rifle, settle down flat on your stomach and wait. You will appreciate a warm coat and possibly even a thermos of tea at this point. Your arrival will have startled the rabbit population and you'll have to wait ten minutes or so for them to return.

Don't have a rush of blood to the head and fire at the first rabbit you see. There will be a number of chances, but some will be too far, or the rabbit might be too young. When you are ready, take the shot, aiming at the head behind the eye if you can. There is a great satisfaction in pulling off a difficult shot over distance. If you are with someone else, never point the gun at them, even if you believe it to be unloaded. An accident at that point could last a lifetime.

In the event that you merely wound the rabbit, you should reload, approach and fire point point-blank at the spot behind the eye. Try to avoid causing unnecessary suffering. If you have missed, either move to another position, or read a book for half an hour. It will take that long for the rabbits to come out again.

Rabbits bleed, so have a plastic bag ready for transport. All you have to do now is skin it and eat it.

Skinning the Rabbit

This is not a difficult process, though it is a little daunting the first time. If you have a heavy-bladed cleaver, simply chop off the four paws. If you are stuck with only a penknife, break the forearm bones with a quick jerk, then cut the skin around the break in a ring. Remove the head in the same way. A serrated edge will cut through the bones, but a standard kitchen knife is likely to be damaged if used as a chopper.

Cut a line down the middle of the chest, from head to anus. This can be fiddly if you're on your own, and a serrated edge is very useful. Be careful not to cut into the abdominal cavity – if you do, the stomach and intestines will spoil the meat.

With fingers, you can now pull back the skin to the hip and shoulders, yanking the fur off the legs like sleeves. When the legs are free, take hold of the fur at the neck and pull downwards. The pelt will come off in one piece, leaving you with the carcass. The belly is quite obvious and bulges with intestines.

Holding the carcass upside down, take a pinch of loose skin near the rear legs and cut a line across it. As you turn the rabbit the right way up, most of the intestines will slide right out immediately and anything that doesn't can be scooped out with ease.

There is a partition between the stomach area and the upper chest that can be broken with a little pressure. Behind it, you will find the heart, lungs and a few other bits and bobs. Pull it all out. The heart, liver and kidneys in particular can be very tasty, but the intestines and stomach should be left well alone.

It is worth taking a moment to have a look at the various inner organs. Male rabbits will have testicles that should also be cut away. This is not for the squeamish, but that is the point of the chapter. If you buy a pork chop, we think you should realise what has gone into providing that meat for you. In a sense, killing for food is a link with ancestors going back to the caves.

Preparing a Meal

This isn't the place for a formal recipe, but it is worth covering the next stage. You could spit-roast the rabbit, but it is easier and more common to joint it – that is, remove the legs by cutting through the joints. Fillets can also be taken from pads of flesh near the spine. You can take a fair amount of meat off a single rabbit – enough, with vegetables, to feed two men and provide hot broth against a winter chill.

Place the meat in a pot with water and bring it to the boil, adding courgettes, a little garlic, carrots, leeks and celery until the pot is half full. Let it simmer for half an hour to forty-five minutes. Wild game is often pretty chewy because the muscles are used much more often than tame animals'. Nevertheless, rabbit cooked in this way is delicious and the broth is very good indeed.

HUNTING AND COOKING A RABBIT

THE SOLAR SYSTEM
(A QUICK REFERENCE GUIDE)

THE SUN. THE CENTRE OF THE SYSTEM

- 93 million miles from Earth (149 million km).

- The Sun alone makes up 98% of all the mass of the solar system. If it was empty, it would take 1.3 million Earths to fill it. The temperature on the surface is a mere 6,000 °C/ (11,000 °F), while the internal temperature is 15 million °C (27 million °F).

- Age: Best current guess is 4.6 billion years. We expect it to survive for another 5 billion years before becoming a red giant, then a white dwarf, before finally burning out. Do not worry about this – the Earth and everything else in the solar system will be destroyed during the red giant stage.

MERCURY

- Mercury is the closest planet to the Sun, at only 36 million miles (57 million km). Second smallest in the system. The surface is cratered in a similar way to Earth's moon. There is a thin atmosphere, containing sodium and potassium from the crust of the planet. Most of Mercury seems to be an iron core.

- Temperature: Hot. 430 °C (810 °F) by day, −180 °C (−290 °F) by night.

- Rotation around Sun (Mercury's year): 87.97 days. This is the fastest in the solar system and as a result, Mercury was named after the Roman messenger to the gods, who had wings on his feet.

- Moons: None.

VENUS

- The second planet from the Sun, at an average 67 million miles (108 million km). Venus has been called the morning or evening star, also Hesperus and Lucifer. Venus is the brightest object in Earth's sky apart from the Sun and our moon.

- Venus can be seen crossing the Sun in 2012. If you miss that one, you'll have to wait until 2117, which is quite a long time. Remember that pinhole or reverse projection from a telescope is a good idea when looking at the Sun – **never** look at it directly, especially with a telescope. The Sun would be the last thing you ever see.

- Rotation around Sun (Venus year): 224.7 days.

- Moons: None.

- Atmosphere: Complete cloud cover resulting from 97% carbon dioxide, the rest nitrogen. Hostile to life as we know it. Surface pressure 96 times that of Earth, so before you could even begin to choke, you'd be squashed flat. The average surface temperature is 482 °C (900 °F). Uncomfortable, to say the least.

- Venus was named after the Roman goddess of love because lonely men sitting in observatories can be quite susceptible to shiny, pretty things in the sky. Its movement across the heavens has nothing to do with actual love, however.

EARTH

- The third planet from the Sun, at 93 million miles (149 million km).

- Like baby bear's porridge, Earth is neither too hot nor too cold. It is just right. Its atmosphere is made of nitrogen, oxygen, 0.03% carbon dioxide and trace gases, such as argon.

- Earth is the fifth largest planet in the system. It has a magnetic field and a liquid nickel-iron core.

- Rotation around the Sun (Earth year): 365.25 days.

- It has an elliptical orbit that means the Sun–Earth distance varies from 91 to 95 million miles at different times. The Earth rotates on the same plane as nearly all of the other planets in the system (except for Pluto), as if they are imbedded in the surface of an invisible plate. Very neat. We call it home.

- Moons: One, which rotates around the Earth in 27.3 days. With an astonishing lack of imagination, we call it 'The Moon'. (This is a bit like the London *Times* calling itself *The Times* because it was first, while all other *Times* newspapers include a city – the *Boston Times*, the *New York Times* and so on.)

MARS

- Fourth planet from the Sun, at an average of 141 million miles (226 million km).

- Gravity: One third that of Earth's.

- No significant magnetic field, which suggests the core is now solid, though it may have been liquid in the past.

- Rotation around Sun: 686.98 days.

- Average temperature: –55 °C (–67 °F).

- Mars has ice caps at both north and south poles, made up of water ice and frozen carbon dioxide. It has an atmosphere of 95% carbon dioxide, 3 % nitrogen and 2% argon and trace gases. Like Earth, it is tilted on its polar axis and experiences seasons, which can involve ferocious dust storms. Despite various probes and landings, we have yet to set foot on the red planet.

- Moons: Two, named Phobos (Fear) and Deimos (Panic). Mars was named after the Roman god of war. The Greek version of Mars was the god, Ares, who had two sons. The moons are named after them.

JUPITER

- The fifth planet from the Sun, at an average 484 million miles (778 million km).

- Jupiter is by far the largest planet in the solar system and the fourth brightest thing in our sky, after the Sun, the Moon and Venus. It takes twelve years to orbit the Sun. It is sometimes called the amateur's planet, because it can be found easily with a basic telescope, or even binoculars.

- We haven't been to Jupiter and we probably never will – so our knowledge is based on observation and the occasional orbiter and probe. Science means we are not blind, however. For example, an effect of gravity is that it causes a passing object to accelerate, which is why you will occasionally see film sequences of spaceships using a 'sling-shot around the Sun' effect. The increase in speed can be measured and compared to other figures we already know. Piece by piece, we build up a picture of a planet – even one where the pressure and gravity is so crushing that we are unlikely ever to ever get a probe down to the surface.

- Jupiter's mass can be predicted from its effect on its moons – 318 times that of Earth. However, if Jupiter were hollow, more than a thousand Earths could fit inside, which means it must be composed of much lighter gaseous elements. This was confirmed by the *Galileo* probe in 1995, which dropped into the outer reaches of the atmosphere and found them composed of helium, hydrogen, ammonia and methane. In many ways, Jupiter is a failed sun – 80 times too small to ignite.

- Beneath the gas layers, pressure increases to more than 3 million Earth atmospheres. At that level, even hydrogen has properties of a metal and Jupiter has a solid core that must be one of the most hostile places imaginable. Winds there will range up to 400 mph and at those pressures, the chemistry of the universe that we think we understand will be completely alien. At temperatures of between –121 and –163 °C (–186 to –261 °F), ammonia will fall as white snow.

- Moons: Around 61, with a faint ring of debris. There are hundreds, perhaps thousands, of rocks orbiting Jupiter. Whether they are referred to as moons or not is a matter of opinion. Galileo discovered the four largest in 1610. They are: Io, Europa, Ganymede and Callisto. Given their size, they deserve a special mention. They are named after lovers of the chief god of the Greeks, Zeus, whom the Romans called Jupiter.

1. **Io.** The closest to Jupiter, pronounced 'eye-oh'. It has a diameter of 1,942 miles (3,125 km), a little less than the Earth's moon. It is intensely volcanic and its closeness to Jupiter's magnetic field generates three million electrical amps that flow into Jupiter's ionosphere. It orbits Jupiter in 1.77 days, at a distance of 220,000 miles (354,000 km).

2. **Europa**. The smoothest object in the solar system. It takes 3.55 days to orbit Jupiter. Its surface is ice, but a weak magnetic field of its own may indicate that there is liquid salt water below the surface. It has a diameter of

just over 1,961 miles (3,155 km). Europa orbits Jupiter at a mean distance of 420,000 miles (670,000 km).

3. **Ganymede.** The largest moon of Jupiter and the largest moon in the solar system, with a diameter of 3,400 miles (5,471 km). It orbits Jupiter at a mean distance of 664,000 miles (1,068,000 km), taking 7.15 Earth days. Ganymede is larger than Mercury.

4. **Callisto.** The last of the Galilean moons. It has a diameter of 3,000 miles (4,828 km) and orbits at 1,170,000 miles (1,880,000 km) from Jupiter. It is similar in size to Mercury and orbits in 16.7 Earth days.

SATURN

- The sixth planet out from the Sun, at 856 million miles (1,377 billion km).

- Like Jupiter, it is a gas planet, with atmospheric pressure condensing hydrogen into liquid and even metal towards the core. Still, we think the overall density would be low enough for Saturn to float on water. It takes 29.5 years to orbit the Sun.

- The atmosphere is composed of 88% hydrogen, 11% helium and traces of methane, ammonia and other gases. Wind speeds on the surface are more than 1000 mph (1,600 kph).

- The rings stretch out more than 84,000 miles (135,000 km) from Saturn's centre. They were first seen by Galileo in 1610, though he described them as handles, as he saw them end on. The Dutch astronomer, Christiaan Huygens was the first to recognise them as rings, separate from the planetary surface.

- Temperature: –130 °C (–202 °F) to –191 °C (–312 °F). (Very cold!)

- Moons: Quite a large number if you count very small pieces of rock, but there are fifteen reasonably sized moons, ranging from Titan, the largest (second only to Ganymede in the solar system and even possessing a thin atmosphere), down to Pan, which is about 12.5 miles (20 km) across. The NASA probe, '*Huygens*' landed on Titan in 2005.

- Saturn is the Roman name for the Greek god, Cronus, who was father to Zeus.

Uranus

- The seventh planet from the Sun, at an average distance of 1.78 *billion* miles. (2.86 billion km).

- It has 11 rings and more than 20 confirmed moons, though as with Saturn and Jupiter, there are probably many more we haven't spotted yet. It is 67 times bigger than Earth, but has a mass only 14.5 times that of Earth, qualifying it for gas giant status, though on a smaller scale to Saturn and Jupiter.

- The space probe *Voyager 2* reached Uranus in 1986, our only source of knowledge at the time of writing, apart from Earth observation.

- Rotation around Sun: 84 Earth years, though it spins on its own axis even faster than Earth – 17.25 hours.

- Uranus has an atmosphere of 83% hydrogen, 15% helium and 2% methane. The planet core is nothing more than rock and ice. It has a huge tilt on its polar axis, so that one pole then the other points at the Sun. This means each pole receives sunlight for 42 Earth years. Average temperature: –197 °C (–323 °F) to –220 °C (–364 °F).

- Moons: 27. All named after Shakespeare characters, with names like: Cordelia (closest), Ophelia, Bianca, Puck, Rosalind, Desdemona, and so on.

- In mythology, Uranus was the father of Saturn, grandfather to Zeus/Jupiter.

Neptune

- The eighth planet from the Sun at 2.8 billion miles (4.5 billion km).

- Neptune is the fourth largest in the system. It has four rings and eleven known moons. It is the last of the gas giant or Jovian planets – seventy-two times Earth's volume, seventeen times its mass.

- It is believed to be composed of ice around a rock core, under an atmosphere of hydrogen, helium and methane.

- Every 248 years, Pluto's erratic orbit brings it inside the 'shell' of Neptune's orbit, making it the furthest planet from the Sun for a twenty-year period. The last time this happened was from 1979 to 1999, when Pluto moved back out. Neptune is the last of those planets that orbit on the same flat plane as Earth.

- The existence of Neptune was predicted before it was seen, like Halley's comet. The orbital track of Uranus seemed to be affected by the gravity of a large mass. The path and location of that mass were mathematically plotted, then searched for – and Neptune was found. It was first observed in 1846.

- The only vehicle from Earth to reach Neptune was *Voyager 2*, in 1989.

- Rotation around the Sun: 164.79 Earth years. It has an axial tilt of 29.6° compared to Earth's 23.5°, suggesting it has a similar movement of seasons, though to be honest, it's so cold, you'd hardly notice, or care.

PLUTO

- The ninth planet from the Sun at 3.65 billion miles (5.87 billion km).

- Pluto is what happens when a stray lump is slowly drawn into a neat solar system. However, Pluto is grown up enough to have a moon and has a tiny effect on the orbits of Neptune and Uranus. It is so small and distant that, even knowing it was there, it still took the telescopes of the world twenty-five years to find it for the first time in 1930. It took until 1978 for anyone to spot the single moon, Charon.

- We haven't managed to get a probe out that far, but the Hubble telescope has mapped 85% of Pluto's surface. It has polar caps and seems to be a ball of rock and dirty ice. It does have a thin atmosphere of nitrogen, carbon dioxide and methane.

- Being a dark and miserable place, Pluto was named after the Roman god of the Underworld (Hades to the Greeks). Charon was the boatman who ferried the souls across the River Styx.

SEDNA – BUT IS IT A PLANET?

- In 2004, Dr Mike Brown of the California Institute of Technology announced the discovery of a tenth planet – one about three quarters the size of Pluto, more than 84 billion miles (134 billion km) away from the Sun. Sedna is reddish coloured, has no moon and its classification as a planet is somewhat dubious. There are, after all, more than a few scientists who think that Pluto should be declassified, never mind this tiny lump of ice. At least Pluto has a moon.

And Finally, Comets, Asteroids
and other debris...

- The Sun is such a massive object that its gravity affects a vast volume of space, trapping objects such as **Halley's Comet**. These tend to be dirty balls of ice, sometimes just a few miles across. Halley's was large enough to have an effect on the orbital paths of the system and Edmund Halley's achievement is that he predicted this mathematically without seeing the comet. In fact, he never saw it. It wasn't until sixteen years after his death, in 1758, that sky watchers on Earth saw the comet once more. It is visible from Earth every 75–9 years and has been recorded since 240 BC. The next appearance is in 2061. It is extremely unlikely that the authors of this book will see it, but there is a chance you will...

- The **inner asteroid belt** lies between the orbits of Mars and Jupiter. It is composed of hundreds of thousands of rocks varying in size from grains to large ones hundreds of miles across. It may be debris from a planet-sized collision, or just the building blocks of the system, left over after everything started cooling.

- **Meteors** reach the system as it travels in space on the end of the milky way galaxy. They are usually made of stone silica, more rarely iron or nickel or a mixture of all three. They can make a bright trail as they reach the Earth's atmosphere and hit friction. If they don't burn up, they can hit the planet below with more force than an atomic bomb – but that almost never happens. (See Dinosaurs.) The best time to look for them is the 9–16 August and the 12–16 December. Meteors in the summer shower are known as the Perseids, as they appear in the constellation of Perseus. At its height, one a minute can be seen. Meteors in the winter meteor shower are known as the Geminids, as they appear in Gemini, near Orion. Both showers should be visible even from urban locations. They won't last for ever – the Geminids only came into existence in 1862.

- That's it. The rest is space and cosmic radiation.

QUESTIONS ABOUT THE WORLD
– PART THREE

1. How do ships sail against the wind?
2. Where does cork come from?
3. What causes the wind?
4. What is chalk?

1. HOW DO SHIPS SAIL AGAINST THE WIND?

When the wind is coming straight at a boat or ship, it would seem impossible to sail into it. It can be done, however, by clever use of the sails and rudder.

Fig. 1 is a plan view of a small boat with the main sail and rudder visible. The wind comes from the direction of the three arrows marked 'a' and would tend to turn the boat in the direction of the arrow 'b'. To counteract this, the rudder is put over as at 'c' and the weight of the water against the rudder pushes the boat in the direction marked 'd'. Between these two forces, like a pip being squeezed, the boat slides ahead in the direction 'e'.

In order to get to a point windward (upwind), the boat must make this manoeuvre first to starboard (right) and

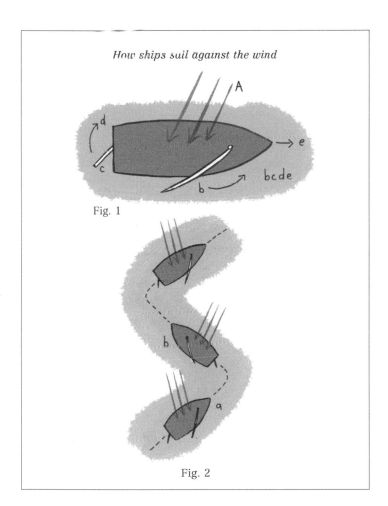

How ships sail against the wind

Fig. 1

Fig. 2

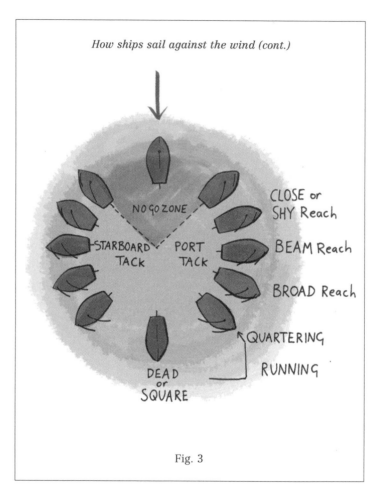

How ships sail against the wind (cont.)

Fig. 3

then to port (left), as in Fig. 2. The boat sails to starboard, 'a', then after a time, the rudder is put over, 'b', and the sail is set over on the other hand – and the boat heels over and progresses on the port 'tack'. By this changing or 'tacking' from port to starboard and vice versa, the boat can, by a zigzag course, reach a point from which the wind is blowing – and it has sailed against the wind.

2. WHERE DOES CORK COME FROM?

When a plant is damaged and its internal tissues are exposed, it is open to fungal and bacterial attack. In a similar way to the human body producing pus, many plant defence mechanisms release a fluid as part of the healing process. From a tissue within the plant, called 'callus', new cells are formed to close the wounded parts. These cells quickly become brown, and assume the nature of true cork.

As man needs cork in rather large quantities, he has found a tree that produces it in large supply, the Cork Oak (*Quercus suber*). The outer bark of this tree largely consists of cork, which can be peeled off in large strips during the summer season. This process injures the tree just enough to encourage it to produce more cork to replace what has been lost. That said, many wine producers are moving over to either plastic corks or screw-top bottles.

3. What causes the wind?

Winds are air currents and their prime cause is temperature differences created by the uneven heating of the Earth's surface and atmosphere by the sun.

Polar regions can be 160 °F colder than equatorial regions. Also relevant is the fact that in the tropics day and night temperatures differ by more than 50 °F. In addition, every mile we rise above sea level will drop the temperature by an average of 17 degrees.

The Earth's rotation also complicates matters, forming regular winds that were a boon to trade in the days of sailing ships. Currents of cold air from the arctic regions cannot keep pace as it spins and are deflected westward. North winds become north-west winds; the same happens in the southern hemisphere, south winds become south-east winds.

4. WHAT IS CHALK?

Chalk is a soft kind of limestone (a carbonate of calcium). It occurs in many parts of England and the ranges of hills in the south-east are composed almost entirely of it. It consists of the shells of tiny animals called 'foraminifera' which live in all parts of the ocean. When the foraminifera die, the insoluble shells form a sludgy deposit which hardens under pressure, and we get chalk. It is often found with flint, another stone composed of fossilised organic remains. Despite its relative softness, chalk can be found in huge cliff formations, such as those on the south coast of England at Dover and Lyme Regis.

TANNING A SKIN

MAKING LEATHER from skins must be one of the oldest human skills. That said, it isn't at all easy to get right and it's worth knowing that small skins (like those of rabbits) can be air-dried after the fat has been cut away. The result will have the stiffness of a bit of cardboard, but there is a very good chance it will feel no better *after* the tanning process. Larger skins have to be tanned, or they simply rot.

First of all, cut away any obvious pouches of flesh on the inner side of the skin. The best way to do this is to stretch the skin onto a board, held in place with tacks at the edges. Use a sharp knife and a lot of care to remove the marbled pink fat without puncturing the skin beneath it. Stone-Age peoples used flints and bones to scrape hides. They also chewed them to make them soft. You might want to try this, though we thought it was going a little too far.

You don't have to get every tiny scrap of fat, but be as thorough as you can. A rabbit skin can be left in a cool room for about ten days and it will dry. Covering it in a heavy layer of salt speeds the process and also helps to prevent any smell of rotting meat. You may want to change the salt after two or three days if it becomes damp or obviously contaminated. When the skin has dried it will be quite rigid. At this point, you could trim off the rough edges with a pair of scissors.

Tanning is the chemical process that makes skin into leather – a waterproof, hard-wearing and extremely versatile material. Almost all the leather we use is from cows, as it is produced as a natural by-product of eating beef. Various chemicals are useful in the process, including traditional ones from boiled brains or excrement. However, we used aluminium potassium sulphate. As well as growing crystals, alum solution can tan skin.

Once the skin is completely dry, it can be dipped in warm water with a little soap to cut the grease. There is a membrane inside all animal skins that must be removed before tanning. One way is to rub the skin back and forth on an edged object, like a wooden board or a large stone. We found wire wool useful, as well as the back of a kitchen knife. It took a long time. It smelled. Peeling off sticky wads of fatty membrane was not an enjoyable experience. Still, no one said it would be easy.

When the skin was about as clean as we could make it, we trimmed one or two of the rougher edges and prepared a solution of 1 lb (453 g) alum, 4 oz (124 g) washing soda (sodium carbonate) and 8 pts (4.55 l) of water. It fizzes as you mix it up, but don't worry.

A rabbit skin should be left in the solution for two days, though larger skins can take up to five. Be careful how you dispose of the liquid, as it's a pretty potent weedkiller and will destroy grass.

When you take it out, it will be sopping wet and the skin side will have gone white. You now need to oil it thoroughly and leave it overnight. Ideally, you would use an animal oil, but those aren't easy to get hold of. We chose linseed oil, which is usually used for cricket bats and church pews. It smells quite pleasant, as well. We placed a plastic bag over it to seal in the oil and left it for another two days.

TANNING A SKIN

The next stage is to let the oiled skin dry, fur side out, but only until it is damp. This stage is crucial to create a soft final skin. Whilst damp, you must 'work' it. This means gently stretching it and running it back and forth over a smooth wooden edge, like the back of a chair or a broomstick held in a vice.

How soft the final version is will depend on how well it was tanned, how much flesh still adhered to the skin membrane and how conscientiously you work it. If it does dry out, and is too stiff, it is all right to dampen it again and repeat the process. It will get softer, but it could take a few sessions.

Finally, a quick dip in unleaded petrol is worth doing, just to clean it and cut through excess oiliness. It will make the skin smell of petrol, obviously, but this fades in a day or two. You really should get an adult to help with this.

Once you have allowed it to dry completely, imperfections can be removed with a sanding block. The skin will resemble waxed baking paper, but it should be *strong* and *flexible* – and it shouldn't smell like a dead animal. Calfskin has been used as paper and rabbit skin could also be written on, though it would serve better as an outer sleeve for a small book, a drawstring pouch or, with a few more, perhaps a pair of gloves

SEVEN MODERN WONDERS
OF THE WORLD

───❖───

THE SEVEN ANCIENT WONDERS are set in stone, but any modern seven must in some sense be a personal choice. Humanity has created many, many wondrous things. A Picasso painting is a wonder, as is a computer, a jewelled Fabergé egg, an aria by Mozart, the motor car, a cloned sheep. The list could be endless.

However, examples such as those don't seem to match the original style and intention of the original ancient wonders. Surely a modern seven should have some echoes in the old ones. Otherwise why have seven, say, and not nine? Our list comes from two rules. 1. It must be man-made, so no waterfalls or mountains. 2. It must take your breath away. Here are seven modern wonders. You cannot look at any of them without this thought: How on earth did we build that?

1. THE CHANNEL TUNNEL

An engineering project to bore a tunnel between Folkestone in Kent and Calais in France – a distance of 31 miles (50 km), with an average depth of 150 ft (45 m) under the seabed. France and Britain used huge boring machines, cutting through chalk to meet in the middle for the first time since the last ice age. When they did meet, there was less than ⅔ inch (2 cm) error, an astonishing feat of accuracy.

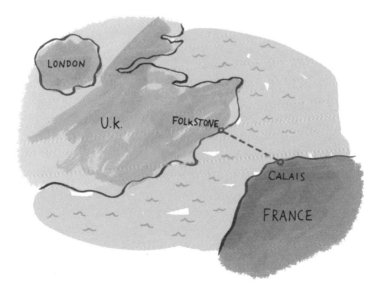

The Channel Tunnel

It took 15,000 workers seven years and cost £10 billion. Part of the structure is a set of huge pistons that can be opened and closed to release the pressure built up by trains rushing along at 100 mph (160 kph). There is also some 300 miles (482 km) of cold-water piping running in the tunnel to ease the heat caused by air friction.

On the British side, the chalk that was dug out was left at Shakespeare Cliff near Folkestone. As a result, more than 90 acres (360,000 m2) were reclaimed from the sea.

2. The Great Wall of China

At 4,000 miles (6,400 km), it is staggering for its sheer size and the effort required to build it. The Great Wall still stands today, though it is obviously not a modern creation. It was begun more than two thousand years ago during the Qin Dynasty. Qin Shi Huang was not a man of small imagination. When he died, he was buried with more than six thousand life-size terracotta warriors and horses.

The Great Wall was designed to keep Mongol invaders out of China, though it failed to stop Genghis Khan. It has a system of watchtowers and forts to protect inner China. Sadly, some sections have collapsed or been destroyed.

The Great Wall of China

SEVEN MODERN WONDERS OF THE WORLD

It is a myth that the Great Wall of China can be seen from the Moon. Many man-made objects can be seen from space at low orbit, like cities, rail lines, even airport runways. From the Moon, however, the Earth looks as if we've never existed.

3. The CN tower in Toronto, Canada

The strange thing is that the CN Tower isn't better known. It is the tallest free-standing man-made structure on earth. To be absolutely fair, its main function is as a television and radio mast and towers just don't catch the imagination in the same way that giant office buildings do. Still, it is 1,815 ft (553.21 m) tall and is a fraction over an inch off perfectly true. It was designed to withstand winds of more than two hundred miles an hour.

4. The Itaipú Dam

This colossal dam stands on the Paraná river on the border of Brazil and Paraguay. To build it, workers removed 50 million tons of earth and stone. The dam itself is as high as a 65-storey building. It used enough

The CN Tower in Toronto

SEVEN MODERN WONDERS OF THE WORLD

concrete for fifteen channel tunnels and enough iron and steel to build 380 Eiffel Towers. By anyone's standards, that is extraordinary.

The hydroelectric power station run by the dam is itself half a mile long. It contains eighteen electric generators, with 160 tons of water a second passing through each one. 72% of Paraguay's total energy consumption comes from this one dam.

The Itaipú Dam

5. THE PANAMA CANAL

One of the reasons the Panama Canal makes it to this list is because it joins two vast oceans and splits two continents. It is 50 miles (80 km) long. Before it was built, a ship travelling from New York to San Francisco would have been forced to go all the way around South America. The canal took almost 8,000 miles (12,800 km) off that journey.

The Panama Canal

It was originally a French project under Ferdinand de Lesseps. Although he was well respected in France, it took all of his charisma and energy to raise the vast capital needed to begin the enterprise. When work did begin, his men had to contend with parasites, spiders, snakes, torrential downpours and flash floods. Far worse was the threat of disease. Yellow fever, dysentery, typhoid, cholera, smallpox and malaria were all common. In those conditions, up to 20,000 workers died. In 1889, De Lesseps' company went bankrupt and the investors lost their money.

In 1902, the American government agreed to take on the Panama Canal and at the same time supported the independence of Panama from Colombia. The American president, Theodore Roosevelt, told his engineers to 'make the dirt fly'. The Americans rebuilt the site and set to work. By 1910, there were 40,000 workers on the canal and Roosevelt had brought in the army. It was completed in 1914.

The principles are similar to any British canal, but each lock gate of the Panama Canal weighs 750 tons. 14,000 ships go through the canal each year.

6. The Akashi-Kaikyo Bridge in Japan, also known as the Pearl Bridge

The longest suspension bridge in the world. It is miles shorter than the longest actual bridge – the Lake Pontchartrain Causeway is over 23 miles (38 km) long – but there is something particularly awe-inspiring about enormous suspension bridges. This one is 6,532 ft (1,991 m) long. It took ten years to build and cost around

The Akashi-Kaikyo Bridge

£2 billion. It is 2,329 ft (710 m) longer than the Golden Gate Bridge in San Francisco.

After a tunnel, a wall, a tower, a dam, a canal and a bridge, most of the truly impressive human building projects have been covered. The last choice may not be physically enormous, but it represents the next stage and the future to come.

7 . THE SPACE SHUTTLE

The Space Shuttle may be the most complicated machine ever built. It is the world's first proper spacecraft and though the current shuttles are approaching the end of their useful lives, they were the first step from single-use rockets to the dreams of science fiction. The first one was actually called *Enterprise* after the *Star Trek* ship, though it was only a test plane and never went into space. It is currently in the Smithsonian museum. Five others followed: *Columbia*, *Challenger*, *Discovery*, *Atlantis* and *Endeavour*. Columbia first went into space in 1981, beginning a new era of space flight. The programme was temporarily suspended in 1986 when *Challenger* exploded shortly after lift-off, killing the seven-person crew.

The Space Shuttle

On re-entry, shuttle skin temperature goes up to 1650 °C (3000 °F). It is the fastest vehicle mankind has ever designed, travelling at speeds of up to 18,000 mph (29,000 kph).

It is used as an all-purpose craft, capable of launching and repairing satellites and docking with the International Space Station in orbit.

BIRD WATCHING

———— ✦ ————

BIRD WATCHING is a popular British hobby but you don't need to sit in a field for hours on end making strange noises to enjoy it. Putting a bird feeder in the garden will do the trick. These can be bought (usually for under £5) in any garden shop. Fill them full of peanuts, seeds or even breakfast cereal and you will be amazed at the variety of birds that will come and visit (not many will turn down a free meal).

Our wild birds face many threats to their survival, mainly due to environmental changes and destruction of their habitats. By putting food out all year round, you can help combat these threats. Food is in short supply from autumn to early spring especially, and the cold weather also makes survival difficult. In the summer months, parent birds will feed from the food you put out and this will allow them to give their offspring food they can find in the wild – like caterpillars. Not only will you be able to enjoy watching but you'll know you are really making a difference to their lives.

Over the next pages are some of the garden visitors you will get.

Blackbird
Turdus merula (10 inches)
Distinctive features: black with a
yellow beak, females are browner.
A striking and well-known bird,
resident all over the British Isles.

Blue Tit
Cyanistes caeruleus (4½ inches)
Distinctive features: bright blue
cap, yellow breast and blue back.
Smaller but more aggressive than
the Great Tit. Loves milk.

Brambling
Fringilla montifringilla (5¾ inches)
Distinctive features: white rump,
reddish breast.
A winter visitor to Great Britain.
Loves the woods, beech trees
in particular.

Bullfinch
Pyrrhula pyrrhula (5¾ inches)
Distinctive features: pinkish breast,
blue back and a white rump.
Looks a bit plump, likes cover,
rather retiring and shy.

Goldfinch
Carduelis carduelis (4¾ inches)
Distinctive features: yellow wing
patterning and a red face.
Feeds low down, has a bouncy
dancing flight.

Great Tit
Parus major (5½ inches)
Distinctive features: black cap,
blue back and yellow flanks.
Loves hedges and gardens. This
one will go after your milk bottles.

Greenfinch
Carduelis chloris (5¾ inches)
Distinctive features: a plump
greenish bird with yellow
wing patches.
Common in parks and gardens.
A resident of Great Britain.

House Sparrow
Passer domesticus (5½ inches)
Distinctive features: brown crown,
white flanks.
British resident, breeds close to
human habitation.

Robin
Erithacus rubecula (5½ inches)
Distinctive features: unmistakable,
bright red breast.
Very territorial: you rarely see two
together. Very proud bearing.

Song Thrush
Turdus philomelos (9 inches)
Distinctive features: whitish breast
with bold spots, brownish back.
Needs cover for nesting but
stays close to human habitation.
A beautiful voice.

Starling
Sturnus vulgaris (8½ inches)
Distinctive features: spotty
black and white with green
shoulder blades.
Spotted from cities to remote coasts,
one of our best-known birds.

Swallow
Hirundo rustica (7½ inches)
Distinctive features: easily
recognisable with its long
V-shaped tail and chestnut throat.
Summer visitor, likes open
country, majestic in flight. Can be
seen all around the British Isles.

Wren
Troglodytes troglodytes (3¾ inches)
Distinctive features: russet brown
plumage with a short, erect tail.
Very active, forages in bins, flies
like a whirring bee.

Yellowhammer
Emberiza citrinella (6½ inches)
Distinctive features: yellow head
and breast.
Feeds on the ground, likes
hedgerows and farm land,
twitches its tail.

SOME TIPS:

1. Don't put too much birdfeed out at any one time so as
    to ensure it stays fresh and dry.

2. Cats love birds! Make sure your feeders are not
    too close to a cat's hiding place. Equally, make sure
    they are close enough to trees in case a Sparrow
    Hawk attacks.

3. If you start feeding, don't stop as birds will come to rely on you.

4. Different birds like different feeds, so the more variety you supply, the better.

5. Birds get thirsty so provide a little fresh water all year round for them to drink and bathe in.

6. Birds can catch diseases so keep bird tables clean and move feeders around to avoid any unhygienic build-ups.

7. Have patience. Birds may take a little time (3-4 weeks) to be confident enough to feed in your garden.

ILLUSTRATIONS

Illustrations on pp 11, 13-16, 41, 44-6, 50, 53-6, 59, 64, 65, 67, 73, 78, 85, 87-92, 107, 110, 115, 117-8, 120, 123, 142-3, 145, 149, 152-4, 156-7, 159, 162, 164, 166, 168, 173-4, 180, 183 © Joy Gosney 2008

ILLUSTRATIONS

ILLUSTRATIONS

DANGEROUS THINGS
I HAVE LEARNT

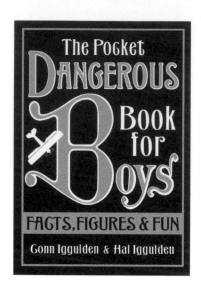

Explorers, spacemen and heroes, great battles and gruesome deaths.

The perfect pocket book of facts, figures and fun for every boy from eight to eighty.

From girls to battles, from anthems to pirates and naval codes.

The perfect pocket book of things to know for every boy from eight to eighty.

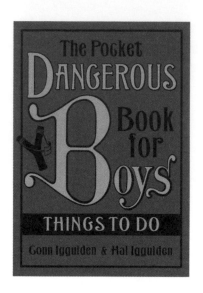

*Build a treehouse, become a conker
champion, master knots.*

The perfect pocket book of things to do for every
boy from eight to eighty.

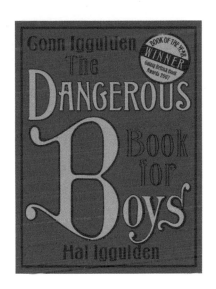

The Original and the Best

The perfect book for every boy
from eight to eighty.

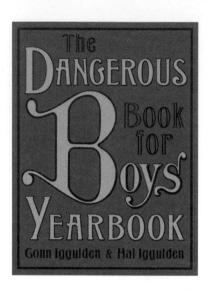

An event for every day, a story for every month.

The perfect yearbook for every boy from eight to eighty.